W9-BIG-617

COMPETING IN TIME

COMPETING IN TIME

Using Telecommunications
for Competitive Advantage

Updated and Expanded

PETER G.W. KEEN

HarperBusiness

A Division of HarperCollins*Publishers*

International Standard Book Number 0-88730-300-5(CL)
0-88730-301-3(PB)

Library of Congress Catalog Card Number: 88-3297

Printed in the United States of America

Library of Congress Cataloging-in-Publication Data

Keen, Peter G. W.
  Competing in time.

  Includes index.
  1. Telecommunication.   2. Business—Communication systems.   3. Business—Data processing.
  I. Title.
  HF5541.T4K44   1988   658.4'5   88-3297
  ISBN 0-88730-300-5
  ISBN 0-88730-301-3 (pbk.)

     92 12 HC 10 9 8

For Lynda with love and appreciation

CONTENTS

COMPETING IN TIME

1 COMPETING IN THE ELECTRONIC MARKETPLACE
The Need to Move Fast

The first edition of this book, published in late 1986, began "Senior managers will learn a lot about telecommunications over the next few years. Many will recognize its growing importance as an opportunity to get significant and sustainable competitive advantages. Others may find out, however, that another firm within their industry has already done so at their expense, or that one outside the industry has used its electronic highway system to intrude into their traditional territory."

The present, new edition, published just over a year later, opens with a stronger version of this still accurate statement: "Senior managers will learn a lot about telecommunications over the next few years. If they don't, they are virtually guaranteeing their firm will be at a competitive disadvantage throughout the late 1980s and early 1990s."

The following list of important business uses of computers and information in telecommunications is not a set of hypotheses but a simple statement of a new competitive reality:

- In financial services the telecommunications network is the franchise; customer data is the product base.

- In manufacturing, computer-integrated manufacturing is a critical survival factor. Without a communications strategy, the firm will be out of the game.

- In distribution, electronic customer service links define the value-added chain.
- In the airline industry, the reservation infrastructure defines the customer relationship opportunity.
- In publishing, the value of data depends on electronic access and distribution.

The lead times for building large-scale communications infrastructures are long; 18 months is a blink of the eye in telecommunications. The companies that have radically changed the marketplace through telecommunications took 5 to 7 years to build the new electronic delivery base. It is hard to catch up when a rival changes the terms of competition in that way.

This is not just technobabble. The last ten years have seen more and more promises of the potential of information technology—computers plus workstations plus data micros. Managers have been offered the Cashless Society, the Information Age, the Paperless Office, and, above all, Productivity.

They have also been bombarded with infohype: evangelical sermons about information technology and competitive advantage that really do sound like "Repent and be converted. The day of judgment is coming!" The sermons use the same cautionary tales: how American Hospital Supply redefined business forever and dominated its industry by building electronic links with its customers and suppliers, how American Airlines climbed to the top of its industry through SABRE, its computer reservation system, how Merrill Lynch took $80 billion of deposits from the banks through its Cash Management Account, etc.

Those stories are true and valid in their main message. They will be repeated in this book. The problem with infohype, though, is that it goes too far beyond what this book aims at doing, which is to alert senior managers to the fact that they can no longer delegate decisions about telecommunications to their technical staff. They must commit time and attention to what is now as much a part of the executive agenda as capital investment plans, human resource planning, and product innovation.

Infohype stresses competitive revolution via technology. This book emphasizes looking for realistic competitive advantage and avoiding competitive disadvantage—the revolution comes from the use of the technology, not the technology itself. Compete in time. The real message of the old stories of American Hospital Supply, American Airlines, and Merrill Lynch and the much more recent ones presented in this book is that when business managers can and do find ways to view telecommunications as a business resource, they harness it in a way that leaves other firms at a sustained disadvantage. This means that no executive can choose to stay ignorant about the topic. Telecommunications is now a necessary strategic capital investment for reasons that range from everyday operational needs to strategic market innovation and renovation. The reasons cover a spectrum:

- Operational necessity—to keep up with the base level of service in one's industry

- Defensive necessity—to protect one's market

- Competitive opportunity—to steal an edge

- Breakaways and preemptive strikes—to change the rules of the games for competitors.

Breakaways are by definition rare and dramatic. Even though some of the most striking examples are almost history, they are still of value in alerting managers to the issue of defensive necessity. They will remain relevant because even though the links between telecommunications and competitive positioning have been clear since 1982, senior managers in most firms recognize that a strategy for telecommunications is important but do not see it as urgent.

The state of management awareness on the subject can be summarized in one sentence: There has been a lot of overselling and naive optimism, together with some real progress, and business continues largely as usual. Senior executives accept that computers are of growing importance to more and more aspects of the firm, but few see telecommunications as a key item on their own planning agenda.

They must. The business implications of telecommunications have become clearer in the 1980s. Telecommunications eliminates barriers of geography and time on service and coordination. It has redefined the base level of customer service in entire industries, changed the economics of market innovation in others, and allowed major companies to reposition themselves. It is a significant product differentiator, especially in mature markets.

There are quite literally hundreds of examples of successful uses of telecommunications to gain and keep a competitive edge. There are also hundreds of instances of woeful failures. The factors that distinguish the winners from the losers in the electronic marketplace are consistent:

Management Awareness.   Senior management is aware of the importance of telecommunications and turns awareness into action. They see it as both important and urgent.

Knowledge of Customer Needs.   The firms that consistently are able to exploit the opportunities of information technology understand customers' actual motivations, not their assumed ones. They also understand the difference between customer research and market research.

Architecture and an Architect.   The architecture is the technical blueprint for the communications highway system and rules of use. The architecture is the strategy. It needs a senior level coordinator.

An Integrated Information Services Function for Integrated Technology.   The skills and careers of telecommunications and information systems, historically separate, must be brought together to provide integrated thinking and skills as communications and computing increasingly merge.

Middle Management Buy-in.   Telecommunications brings radical changes in jobs, work, relationships, and skills. Firms frequently fail to implement a good business and technical strategy because they overlook the vital importance of moving the culture with the strategy.

Seven-year Horizon and Follow-through.   Success in communications planning means being there when demand takes place. Since the lead times are two to seven years, one year "strategic" plans are mainly of value for intellectual merriment. Think ahead. Compete in time.

Managerial Courage.   If innovation were easy, everybody would be doing it. If innovation could be cost justified and guaranteed, there would be no risk. If telecommunications is just about technical operations, business managers can keep out. It isn't easy. It is risky. It needs boldness, real commitment, and business leadership.

This book addresses all these factors, but its central topic is the first one: Senior management awareness and turning awareness into action stand out as absolutely vital in using telecommunications to create business advantage and avoid disadvantage.

1 BUSINESS INNOVATION THROUGH TELECOMMUNICATIONS
The Management Opportunity and Responsibility

When do we lead, when do we follow the leaders in using new technology or applying existing technology in new ways? This question becomes critical when technical risk becomes business risk, and vice versa. Investing in telecommunications is inherently a business gamble. The technology is expensive, rapidly changing, and complex. The stakes increase as traditional marketplaces are changed by electronic delivery, as new markets open up, as some firms use communications successfully to reposition their business, and as others succeed only in creating expensive write-offs on the income statement. When a firm leads, it is betting on an unproven business idea and often on unstable technology. When it follows, it is betting that it will be able to catch up to the leaders.

Given the past gaps between promise and reality in such areas as home banking, office automation, expert systems, debit cards, videotex, videoconferencing, or on-line information services, most senior managers are skeptical about geewhiz technology. They prefer to wait a little and let someone else prove the market is there. What can they do, though, when a geewhiz business idea, based on proven technology, turns out to be a winner? The technology creates a barrier to imitation. Since catch-up time is measured in years, the firms that lack the telecommunications base are locked out.

"When do we lead, when do we follow?" can be rephrased as "What telecommunications infrastructure do we need in order to respond to opportunities and lead the competition, or if we must follow, in order to follow fast?"

THE COST OF FOLLOWING TOO FAR BEHIND

There is ample evidence from many industries that following is actually more risky than at least keeping very close to the leaders.

Banking. In banking, not being able to provide automated teller machines (ATMs) is like not giving customers checkbooks; yet, as late as 1982, many banks were not convinced that ATMs would be profitable. In international corporate banking, leading firms held back from following the lead of Citibank in building a global electronic delivery base. They saw cash management, electronic funds transfer, and foreign exchange trading as separate services. As they automated them, they built independent and incompatible communications systems, instead of focusing on creating the multiservice system onto which a growing variety of services could be added at low incremental cost. They are now wishing they had had a master plan and blueprint earlier.

A June 1987 report from McKinsey and Company and Salomon Brothers points out that "to compete and thrive during the remainder of this decade and throughout the 1990s, banks must integrate the various elements of electronic banking." Citibank is far from playing the hare in the banking race and much of its recent dominance of its industry reflects its willingness to experiment and take risks in the 1970s. The report said that Citibank's 1985 expenditures of $900 million far outpaced runner-up Bank of America, whose estimated expenditures were approximately $500 million—much of which was catch-up expense.

While a large information technology budget helps, several smaller banks have increased profits by forging ahead of the pack. Examples are Banc One Corporation with its credit and cash management account processing, First Wachovia Corporation with its student loan processing, CoreState Financial with its ATMs and point-of-sale terminals, and Bank of New York with its mutual funds. The main element of success in each case was having the telecommunications highway on which to put various applications in place ahead of the competition.

Airlines.   Delta Airlines, for decades one of the best managed airline companies, admitted in 1984, when the federal government proposed to bring lawsuits against American Airlines and United Airlines, that it was now at a competitive disadvantage from not having invested in reservation systems ten years earlier. The suits argued that the other two companies' telecommunications capabilities gave them a dominant edge.

Any airline or travel agent that has not invested in computerized reservation systems and distribution systems or that has to rely on competitors' systems for the sale of their own products faces serious disadvantages in the travel market. Luke Mayhew, head of distribution and marketing services for British Airways, in discussing airline reservation systems as the base for marketing and managing distribution channels, particularly travel agents, says that "it can be argued that the changes in the structure of the industry are largely being dictated by distribution systems and how they are being used by the airlines that have them. The have nots have been faced with real problems" (*Financial Times*, August 5, 1987.)

American Airlines has long set the pace for innovation in the industry. It was the first company to launch a frequent flier program. The purpose of the very popular "AAdvantage" program was not just to gain short-term market share but to accumulate data on the fewer than 400,000 people who accounted for about 70 percent of full-fare travel. Several years earlier, when it was going through difficult times, American began investing in its reservation system SABRE (Semi-Automated Business Research Environment). AAdvantage depends on SABRE.

American Airlines' competitors had to follow its lead and offer their own frequent flier programs. They got all the costs of administration of the program, while American has the customer relationship data, accessed and processed through its SABRE reservation system, with direct links into other such systems. Today SABRE is clearly the market share leader. With 50,000 terminals in 14,000 travel agency locations, the system handles 45 million individual transactions per day. More important though, and a factor that American's competitors now view as crucial, is that the ownership of the world's leading reservation system does a good deal more for American than simply boost revenues. It has set the pace for all other major airlines in terms of marketing, building customer loyalty, and even making acquisitions and mergers. American Airlines' initiative, an apparently technical project, was a springboard to business innovation. The company made almost twice as much profit in 1986 (around $400 million) from SABRE as it did from its airline operations.

Supermarkets and Retailers. Supermarkets have turned their cash registers into point-of-sale banking machines, often at the expense of banks, since many retailers have more effective telecommunications systems than the banks do and have shown more imagination and boldness in using them. The Publix chain in Florida completely pre-empted the state's major banks in electronic funds transfer at point of sale (EFT/POS). By providing the highway for the development of the ATM card and credit card networks, telecommunications thus breaks down barriers between industries. It even redefines them: Is EFT/POS banking or retailing? Who covers the cost of the hardware and the cost of processing? More important, who controls the information being exchanged? Sears, Publix, and Giant Foods (also in Florida) are now powers in banking because they have discovered that the movement of money electronically can be done by anyone, not just financial institutions.

In August 1987 Sears could offer its customers a full range of financial services—from Discover credit cards and Allstate insurance policies to home mortgages, car loans, and Dean Witter mutual funds. The jury is still out on the long-term success of Sears' strategy to reposition its business, again using telecommunications and electronic online service as the springboard, but in the meantime Sears is certainly no longer just a department store.

New industries emerge through electronic delivery. Who controls which parts of the infrastructure? When should firms cooperate and when should they compete? The Florida banks assumed that their control of the payments system would allow them to set the pace and path of change. Publix bet on its ownership of the customer contact point and of the telecommunications highway. It was able to force the main banks to cooperate with it, not the other way around.

Will credit card providers like American Express control more aspects of business air travel because the card establishes the customer relationship and because these companies have access to the airline reservation systems? Should airlines cooperate with banks and credit card firms in integrating reservations and payments and adding new customer service? Should banks compete with each other to establish premium electronic services or should they share resources to create barriers against nonbanking competition? There are no easy answers to such questions, but it is vital for any major firm to recognize when they have to be asked.

Manufacturing.   In the late 1970s Chrysler Corporation was given only months to live by most automotive industry analysts. The role of Lee Iaccoca in leading Chrysler back from near bankruptcy is very well known. Less familiar is the role of telecommunications to support the business leadership. Despite the dismal outlook, the company's planners decided to invest millions of dollars that they could scarcely afford on sophisticated communications and computer systems. Robert Brauburger, chief engineer at the Technical Computer Center stated that "from 1979 to 1983, the company spent the money on technology when they weren't spending it on anything else." Today, largely because of the strategic move toward computer-integrated manufacturing (CIM), Chrysler is out of the hole and in a competitive position beside industry leaders Ford and GM.

Chrysler's vice chairman, Gerald Greenwald, ascribed much of the firm's comeback to its use of communications and computers, which eliminated 20 percent of the company's paperwork and allowed drastic cuts in inventories. With its new ability to build to order, Chrysler no longer needs an inventory of $700 million of unsold cars in 40 assembly plants. Dealers' orders are received and their credit checked electronically. Currently, more than 3,000 dealers are linked to the corporate information system.

Chrysler's Dynamic Inventory Analysis System (DIAS) network links all suppliers to all Chrysler plants and all plants to all markets. Components are always on hand. Assembly line stoppages due to shortage of supplies have been eliminated. In the Canadian minivan plant, terminals linked to manufacturing equipment provide feedback to control centers, and they have cut days or even weeks from the time needed to respond to needs for maintenance and troubleshooting. This is quite a feat when one considers that the DIAS system has to coordinate the flow of producing 12,000 vehicles a day, using approximately 40,000 types of parts and 10,000 employees.

The essence of CIM is the control of an entire factory production system through computers possessing a common database of information about the factory, its capabilities, and its products. CIM enables firms to produce multiple products efficiently, respond to rapid market changes, adapt to shorter product life cycles, develop high-quality custom designs, and drastically reduce inventory. At a time when markets are continually changing, this advanced method of manufacturing will almost surely become the base for U.S. manufacturers—not just in the automotive industry—to lead the competitive response to Asian competition by using information technology—especially CIM and electronic data interchange (EDI) to offset disadvantages of labor costs.

Large Companies. Many large firms have realized that once they have established a network, the cost of adding new traffic to it can be small.

Ford decided that it was better positioned to handle electronic cash management than were banks and applied (unsuccessfully) to join the Society for Worldwide Interbank Financial Telecommunications (SWIFT), the international banks' telecommunications system for funds transfer.

Texaco realized that it could bypass banks for foreign exchange and treasury management activities if it could link into the international dealing networks.

A number of car manufacturers saw that they could use their telecommunications system to allow central staff to get up-to-date information on dealer inventories, yesterday's sales, and so on, to coordinate production and reduce information "float"—the gap between something happening in the field and information reaching the decisionmakers at the center.

British Petroleum found that it could use the banks' technology to manage its own financial transactions and save on commission. In 1986 British Petroleum Finance International managed the company's $4 billion worth of liquid resources and contributed over $40 million to its annual income. In other words it used telecommunications so that it could become a bank. Volvo has done the same and at least a dozen major multinational corporations are close to doing so.

Telecommunications eliminates the dichotomy between centralization and decentralization. Several companies have been able to exploit this to reduce levels of key inventories (including the cash inventory) by around 20 percent. Bechtel, through the imaginative combination of visionary business thinking and visionary technical thinking, is getting ready to field "offshore" project teams and engineering skills in the same way manufacturers set up offshore production. Hewlett Packard has changed the relation between its field sales team and branch offices. The sales representatives use portable personal computers linked to central information stores.

The examples could be multiplied, but the ones just mentioned contain the main messages, that electronic delivery has become the norm in industry after industry; firms have to have the highway system in place before they can carry the traffic.

Suppliers put a terminal in the customer's office, providing a direct link to their computers for placing orders and getting information. The customer gets a new level of service and responsiveness. In any mature industry where it is hard to differentiate the product through price, promotion, or manufacturing, such service becomes the differentiator.

There can be no better service than one that gives direct and immediate access, even outside normal business hours and across time zones. Electronic delivery can also mean more, not less, personal service; the sales force does not have to spend so much time on administration, order taking, and answering routine queries. The terminal does that, and the sales staff can concentrate on selling and face-to-face service. The customer does not have to wait for the salesman or phone the company to get information it needs now. The customer service unit is in effect in the workstation; it is as if the supplier has a branch on your premises.

Getting the electronic delivery base in place ahead of the competition can give a company the advantage of occupancy; customers want only one terminal, which can soon mean one main supplier. Ownership of the customer contact point, such as the airlines reservation terminal, point-of-sale register, cash management workstation, or order entry terminal, provides the base for delivering nontraditional services and capturing the customer relationship.

The evidence is that it is better to get there first with a good enough service than second with a better one, largely because of what Michael Porter calls the "cost of switching." American Airlines does not have the best reservation system, in terms of technical design. TWA's is far better. Chase's cash management services are technically superior to Citibank's.

The occupant can create a new product stream once the system is in place. When the dealer order entry system is in place, it costs little extra to use it for processing warranty claims, distributing updates to price lists, or "downloading" daily reports to the terminal at 2:00 A.M. when the supplier has processed today's transactions. When the customer opens up for business in the morning, the transactions are immediately available. The leader in installing a corporate cash management service can add portfolio management and payroll services to the same base at minimal incremental cost, as has the Royal Bank of Canada, which thus became the de facto pacesetter in its market.

These are not advanced services but already relatively standard requirements for maintaining competitive position among pharmaceutical distributors, corporate banks, car manufacturers, airlines serving travel agents, and insurance firms. Their strategic business options expand or contract depending on the scope and quality of their highway system. The early 1980s showed many examples of this, with Citibank, Thomson Holidays of the United Kingdom, American Hospital Supply, McKesson, and American Airlines being the most widely cited instances. These firms were early winners in the electronic marketplace, not because of the technical quality of their systems, but because of the recognition by the leadership that telecommunications was not just an operational necessity but rather the springboard for business innovation.

In the late 1980s opportunities to create striking market leads are fewer. American Hospital Supply is a legend for showing how being the first to lock in customers by electronic delivery could preempt an entire industry. Now, every firm recognizes the need. American Airlines still holds the industry preeminence opened up by its early investment in SABRE, but now no airline overlooks the importance of reservation systems.

Thomson Holidays of Britain developed a cheap and simple videotex system that enabled the agent and prospective client to become involved in the holiday booking conversation as the videotex TV screen interacts with the Thomson computer center. Thomson's system has become the de facto standard of the European travel trade, and although the other big operations have introduced me-too systems, Thomson's system remains the favorite. The rest of the pack, though, is now back in the game. Thomson is unlikely to find such easy pickings in the near future. That will be true for American Airlines and Citibank, too.

There are, though, still plenty of gaps to fill. The *Wall Street Journal*'s 1987 Technology Review scorecard shows that even now many large firms have not recognized that telecommunications is no longer about technology but about business capital. The *Journal* gave the telecommunications industry a C − for failing to sell its impressive technological know-how to the customer. Other industries got equally low grades because they did not make use of technological advances in a timely manner. "Banks don't put money into where things are going, they only put money into where they are," says Banc One Corporation Senior Vice President John F. Fisher, who is considered one of the industry's best practical futurists. (Banc One has created a major market niche as a provider of electronic services to other financial institutions. It was Banc One's Visa processing capabilities that provided the service base for Merrill Lynch's Cash Management Account (CMA)). The long-term competitive cost of such a view will lead to often unresolvable trouble for many organizations because the cost of following is increasing, not decreasing.

MOVING FAST IN TELECOMMUNICATIONS: SENIOR MANAGEMENT ACTION

Of course, telecommunications does not create business strategy. It opens up new possibilities for it. It also creates new risks. The technology of telecommunications is changing rapidly and involves expensive gambles. The business gambles are even bigger. For every success in innovation in electronic delivery and in moving into new markets through telecommunications, there has been at least one failure of comparable size.

Dun and Bradstreet successfully repositioned itself to become an electronic publisher. The *New York Times* tried earlier and lost. Merrill Lynch's Cash Management Account, which relied on the telecommunications processing base, took away billions of dollars of deposits from banks, but the firm got into trouble in its traditional business, and it is not clear how profitable the venture turned out to be. Retailers, including Sears, which is a leader in the move toward creating the financial supermarket through technology, have had a number of expensive failures with point-of-sale systems over the past 20 years. While electronic cash management is now a proven success, home banking is still a solution searching for a problem. Continental Illinois was one of the most effective innovators in electronic banking and the use of the office technology, but when a firm makes major mistakes in its core activities as did Continental, no amount of technology can rescue it.

Even when the strategy is clear, it is not guaranteed to work. Soon after United Airlines renamed itself Allegis, its chief executive officer,

Richard Ferris, lost the support of first his pilots, then Wall Street, and then his own organization. He is gone and his strategy of vertical integration of travel services via telecommunications has collapsed. This dismantling of Ferris's grand plan may well turn out to be the way the airline business is moving—and the pilots' perspective an error—but when the CEO cannot communicate the vision and get the buy-in of his people, the strategy is irrelevant.

One reminder that innovation via technology is inherently risky comes from Federal Express. Its brilliant use of cellular radio in the delivery van and coordination via communications was a major competitive differentiator around the same time its Zapmail service was a multimillion dollar flop. The project was bold and made good sense, but the demand was missing and an alternative technology of low-cost, high-speed facsimile machines was developing faster than was realized. Perhaps Federal Express should have waited.

The cautionary tales could be multiplied: Citibank and McGraw-Hill's joint venture, GEM; the consortium of CBS, Sears, and IBM losing over $30 million each in a videotex venture; and on the vendor side, US Sprint's loss in 1987 of $1 billion trying to buy market share via advanced telecommunications technology.

Nonetheless, whether viewed in terms of opportunity or necessity, no doubt exists now that telecommunications and business innovation are interlinked. Senior managers simply must understand the linkage and be ready to act. Only they, not their technical staff, can decide when to lead and when to follow: when to make the business act of faith needed to build an infrastructure that will take at least 2 years and probably closer to 10 to provide bottom-line benefits and when to wait until others have clarified the direction to move and reduce the business and technical gambles.

In the 1970s it generally made more sense to follow than to lead. The cost and problems involved in introducing new systems were too often higher than the likely payoffs from pioneering in new business territory. The risks now are in the other direction and at the very least firms need to ask what is the communications base they need for defensive necessity, if not for competitive opportunity. The corporate network is the highway system for the computer traffic of the late 1980s and the 1990s. When that traffic is the firm's business revenue, customer image, and base for growth, it is time to move fast.

THE THREE STAGES OF TELECOMMUNICATIONS MANAGEMENT

Historically, telecommunications management has moved through three stages:

1. The operations era
2. The internal utility
3. The coordinated business resource.

Each of these entails a different mandate, experience and skill base, technology, and managerial perspective on the use of telecommunications.

Telecommunications used to mean telephones, essential to business efficiency but hardly something senior managers needed to factor into their strategic planning. Early data communications by computer were built on the telephone system. The main need in the 1970s was to link terminals to computers; airline reservation systems were among the first widespread applications. The transmission speeds were low and costs high.

In the early 1980s there were rapid changes in every aspect of telecommunications. The plain old telephone system (POTS) could not provide the capacity, speeds, and efficiency needed for computer communications. Satellites and fiber optics communications have transformed the economics and accessibility of telecommunications. Deregulation in the United States, liberalization in the United Kingdom, and protection of its monopoly by most other countries' telephone authority have made communications a volatile political issue. So, too, has the question of "standards." Each major computer and office equipment vendor, such as IBM, Wang, Xerox, and Digital Equipment, had its own proprietary telecommunications systems, which meant that customers had proliferating facilities, incompatible equipment, and escalating cost.

International committees tried to define comprehensive standards while the vendors fought to establish their own. The technology became more and more diverse and more and more complex. There were so many major improvements in voice communications (computerized telephone systems) at the same time as in data communications and in computer hardware and software that the telecommunications manager's job became like trying to change the tires on a moving car.

In the 1980s instability has been the norm in just about every aspect of telecommunications. It has not been easy for companies to coordinate the many technical and managerial issues involved in running phone operations, upgrading them to exploit the immense improvements in voice technology, dealing with deregulation, handling cost-effective data communications for specific applications, trying to eliminate incompatiblities and establish standards, and evaluating emerging technologies and new vendors.

Not surprisingly, the focus has been on managing operations and on technical issues and costs. It has been hard enough for firms to locate and retain good telecommunication specialists and even harder to find generalists who understand both the technology and the business context. As a result, there has often been far too little dialogue between the groups who need to work together, and think together, to create a corporate business resource.

There is a knowledge gap even among the specialists. Many telephone managers have little understanding of data communications, computer staff of telecommunications in general, and data communications people of data processing. With the gaps in knowledge come splits in responsibility. A priority in organizing for telecommunications must be to bridge the gaps.

The Operations Era Management

The operations era was concerned with providing reliable telephone and telex services inside the firm. The skill base was fairly low; supervisors came up through the ranks and were mainly required to have strong operational experience. There was no need for technical expertise, since external suppliers, mainly the phone company, provided technical support.

During this period, central planning was rare. Business units might each have their own telephone manager plus a telex operation. In many geographically dispersed firms, the consequence of this lingers on. Responsibility for communications is scattered among a number of people and budgets. Voice communications are mainly handled on a local basis. The data processing unit eventually took over most aspects of data communications, but the scattered fiefdoms remain. Quite often, senior management has no idea of just how much the firm is spending on telecommunications because of this lack of coordination.

The operations era of telecommunications management is past. It has left behind some old-timers who may hinder change because they think entirely in terms of physical operations. They are not up to date about the new technologies, though they tend to know a lot about a few vendors' products or specific equipment. For them, telecommunications planning means equipment selection, which many of them are only now recognizing. Although it is easy to dismiss the old-time operations managers' experience as irrelevant to the new era, they do know how to make things work. Telecommunications is at the same time very abstract and very concrete. People take for granted that the telephone system works. The operations manager views it in terms of the physical devices and is very aware of how much lies behind the phone receiver and how much has to be done to keep it working.

The operations mindset is still needed for data communications, to support, not substitute for more adventurous planning. Conceptually, for instance, installing a "local area network" to link up personal computers in an office building is relatively simple and the technology proven; however, a myriad of small details must be dealt with: problems of cable size, connectors, power supplies, synchronization, etc. No one—not business managers, data communications gurus, or telecommunications planners (or writers of books on telecommunications and business strategy)—should ever forget that in moving from strategy to cables and boxes, from abstraction to physical system, if the boxes do not work, the strategy is irrelevant.

The Internal Telecommunications Utility

Most large firms began moving to consolidate communications and create an internal utility around 1982. Many companies, both large and small, still handle telecommunications in this way. They view it as a technical function and focus their attention on providing facilities as needed and on controlling costs. The utility reflects a largely reactive strategy.

In the early 1980s the tidy world of the operations era was disturbed by a broad range of new factors. They included:

- Management's concerns over ever-increasing phone costs

- The proliferation of new vendors and products in what had been a very stable industry dominated by AT&T

- Disruptions in service created by the Bell System divestiture

- The growing importance of data communications and computers, and the proliferation of equipment and facilities they create

- The need to plan for computer communications as well as voice.

The message from top management was "Get this under control."

The priorities in managing the utility became to reduce costs and provide some degree of central oversight and planning, often in an environment of highly decentralized operations. This was done mainly by setting up a new bureaucracy. The utility coordinated communications planning through the budgeting and internal accounting process. It controlled costs by tracking them and charging them out.

The new technology base the utility has had to bring together is very diverse. It includes an emerging generation of advanced telephone facilities, satellites for voice and data, specialized equipment for improving the performance and throughput of high-speed, high-capacity transmission links, local area networks for personal computers, automated network management equipment, and entirely new providers of basic and advanced transmission services. New skills have had to be added, too often with difficulty.

Around 1980 the very basics of the technology began to change. Historically, telecommunications relied on "analog" transmission, by which signals were sent along wires with electrical pulses reproducing the waveform of the speaker's voice. Computer data is "digital" in form, meaning that every number, letter, or special code is coded entirely in 0's and 1's. Sending digital information in analog form became increasingly inefficient. Digital transmission was essential to exploit new advances in equipment and high-speed transmission.

By 1985 it was clear that phone calls should also be transmitted in digital form, and that at some stage all information—telephone calls, computer data, pictures, documents, and diagrams—could share the same highway system instead of needing separate and incompatible facilities. The term "integration" became a standard part of the telecommunications vocabulary, and even though most firms were very cautious about moving too quickly when the new technology was unproven, the planning issues for telecommunications had to focus more and more on defining and building a coordinated infrastructure.

Fiber optics and satellites created new economics of scale. The importance of terminals, workstations, and personal computers, especially for customer service, reinforced the need to avoid incompatibility and to share resources. In the United States and in every developed country, the providers of telephone services started planning to shift to data communications and then, by the year 2000, to integrated digital networks.

All these trends changed the stakes, scale, cost, and complexity of managing telecommunications. They left the utility manager with some continuing and often unresolvable problems:

- How to balance short-term cost control against the long lead times involved in anticipating unpredictable increases in demand

- How to impose coherence on the existing multiple voice and data facilities and vendor-specific, and thus incompatible, equipment and networks, while maintaining quality of operations

- How to organize for a very new technical and managerial environment

- How to accommodate emerging needs such as links to customers' computers, high-speed applications like videoconferencing or CIM, and interconnection to other networks (links between retailers and banks, for example).

The telecommunications utility is complex to plan, implement, operate, and advance. It is hardly surprising that technical and cost issues dominate the communications manager's attention in the utility stage or that business managers have very limited insight into what telecommunications is and means. That is why so may companies have been caught unawares by innovators that use telecommunications as a part of an ambitious business strategy or why in almost every industry a few businesses have stolen the march on their competitors. The companies left in the dust by innovators were still in the utility stage, where the telecommunications department or function responds to business needs but does not initiate business strategy.

The Coordinated Business Resource

The transition to the third stage, the coordinated business resource, depends above all on management enlightenment: the recognition at the top of the organization that the topic is no longer an internal utility that is an expensive subset of the firm's administration and operations but about a function that is part of the company's business infrastructure. The recognition is generally stimulated by market pressures or opportunities. These lead to a rapid succession of new applications that rely on data communications: using office technology as a cornerstone for productivity drives, putting a terminal in the customer's office, or linking personal computers to computer databases.

The old utility structure can rarely handle the new demands and the saliency of what has been, up to now, largely a back room, technical function. Senior managers do not start to recognize the importance of telecommunications to key business functions, unfortunately, until lack of facilities, unreliable operations, or the network's "crashing" translate into customers not being able to use an on-line electronic service.

Any managers who have ever planned to get cash at the airport using an automated teller machine at 11:00 P.M. and found the system was "down" appreciate their own customers' feeling and the less than lyrical phrases they use to describe the company when there are problems with the dealer order entry system, the international corporate cash management network, or the advanced digital transatlantic phone network. Telecommunications operations become a business issue when the business is on-line. In fact, it is the firm's cash flow, quality of customer service, and image that are on-line and at risk.

When senior business managers become aware that telecommunications is now important to business effectiveness and not just to operational efficiency, they still have little idea of what to do. It is hard enough for the telecommunications utility manager to handle the technical demands. How can a marketing manager, head of a product division, chief executive officer, or corporate planner begin to think imaginatively and realistically about telecommunications as a business opportunity and then contribute to telecommunications planning?

Creating an organizational strategy for telecommunications requires a new style of business thinking among senior managers, who must also have some insights into key aspects of the technology itself. The question is where to start.

2 PLANNING—FROM VISION TO POLICY TO ARCHITECTURE
The Senior Management Agenda

The place to start creating an organizational strategy for telecommunications is with the business opportunity, not the technology. What aspects of the firm's business vision, its cost of doing business, competitive and market trends, and customer preferences raise telecommunications from being tactical and technical to becoming a strategic business issue? There is no basis for a telecommunications strategy until this question is answered.

Treating communications simply as a utility rarely if ever raises the questions. Decisions are made, mainly in reaction to cost and operational needs. If in fact there is no business reason for committing to a comprehensive electronic highway system, then telecommunications should remain a technical utility, and existing procedures and policies in such areas as funding and business justification will require little revision. There will also be little need for a long-term technical blueprint for the telecommunications base; facilities can be added as required and will mainly involve upgrading equipment and adding transmission lines.

If, however, the business context and market trends and opportunities suggest that the success of the firm may be hampered or facilitated by the scale and design of its communications infrastructure, then important decisions need to be made. The firm's leaders have to take a fresh look at the opportunities of the future. They need vision to create or change the company's policy regarding telecommunications. If the firm is to build a long-term coordinated resource, clear lines of authority and responsibility must be established, as well as procedures for funding a shared corporate infrastructure. An "architecture" is needed. This is a flexible master plan for evolving an integrated capability. It is the base for the ongoing technical strategy that supports, reflects, and even helps shape the business strategy. Such a plan pulls together the many and messy components of telecommunications and computers into a coherent framework. The architecture is the strategy.

These are the components for managing telecommunications as a business resource. The planning sequence—business vision to senior management policy to technical architecture—rests on the quality of the dialogue between business managers and technical personnel or advisers. Two monologues do not make a dialogue. Building the business vision is obviously the domain of business managers and corporate staff. The technical planners' responsibility is to define the architecture. Policy links vision and architecture.

VISION

It is not enough for senior managers to recognize that, say, dealer order entry systems or electronic mail are a useful and natural extension of on-line computer systems or that the firm needs to upgrade its data networks and telephone systems to be competitive. That merely says, "We need a bigger utility." There has to be a broader vision of the business future.

The vision is a picture of the future, an easily understood statement about a practical and desirable, even if not fully predictable, goal. It answers such questions as:

What is happening in our industry and around its edges?

- Where are the new sources of innovation and competitive advantage?
- Where does "service" give an edge?
- What are the leaders doing and why?
- What does all this mean for the basic direction and priorities of our business?

What makes us special? What must we do to remain so?

- What is the core of our business?

- How do we maintain differentiation?
- Why will customers give us their loyalty or give it to others?

Who is our competition?

- Who is our competition now?
- Who will our competition be five years from now?
- What caused the change?
- What pressures does this create for us?
- What opportunities does it open up for us?

What style of organization do we want to be?

- What is our culture and how do we keep it vital and effective?
- How do we want our customers and competitors to see us?
- What makes it easy to work here and what makes it hard?
- What do we mean by "productivity" and how would we recognize it if we saw it?

How can we run our business better?

- Should we improve coordination and responsiveness and combine decentralization with centralization?
- Should we eliminate delays in getting needed information?
- Should we speed up decision making?
- Should we eliminate such problems as service delays, reaching customers, or answering queries?

At first glance, these questions have nothing to do with telecommunications. If, though, the answers relate to it, then a technical utility is no longer being discussed, and the firm will need to rethink the policies and management process for communications.

POLICY

Policy is the small set of mandates and directives from the top of the firm summarized by the ground rule "This is how we do things around here." It addresses questions of authority and accountability. It makes no sense to assign increasing responsibility for turning telecommunications into a coordinated resource and to define standards that must be followed by all units when the authority is unclear or really rests in the business groups. If they choose to ignore the standards, what value are they?

One of the most difficult and common problems in moving from the utility to the coordinated resource is that the responsibilities are increased but the policies and locus of authority remain, often by default, those of the utility. The contradictions and ambiguities this creates are the direct result of management's abdication of leadership. The policy agenda is a key to matching business opportunity to feasible action.

Several policy questions have to be answered explicitly and not by default.

- What pace and degree of change—business, organizational, and technical—are we ready to accept?

- What is our planning horizon?

- How should the business case for building the long-term infrastructure be made? How should it be funded?

- What range of future business services should be anticipated in the technical architecture?

- What are the criteria for selecting vendors, given the relative importance of integrating some, most or all the components of the information technology base?

- Who defines, coordinates, and monitors standards?

Policy is not the same as planning. It sets the criteria for planning and establishes the ground rules for it. Both are needed to define the architecture that is the framework for using the technology.

ARCHITECTURE: THE STRATEGIC TECHNICAL BLUEPRINT

From the senior manager's viewpoint, the architecture is the technical strategy. It is roughly like a city plan, a mixture of fixed main routes, zoning regulations and ordinances, and procedures for extending and modifying existing buildings and expanding or adding roads. The higher layers specify terms and standards for buildings. They break down into lower ones, which include, for instance, the specification of electrical equipment.

A simple example of a corporate telecommunications network is shown in Figure 2–1. It looks far tidier than do real-world large-scale networks but gives some sense of how complex they can be and how

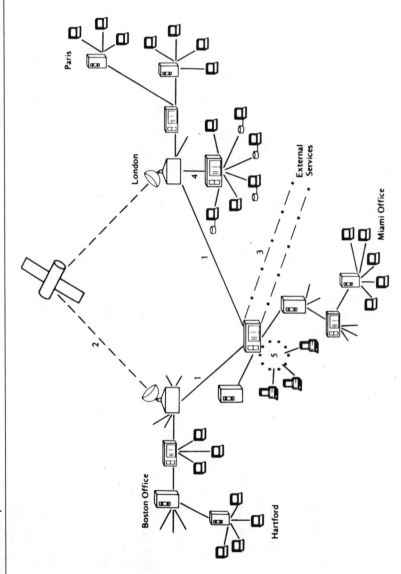

Figure 2–1. Components of a Telecommunications Resource.

Figure 2–1.   continued.

 computer "hosts"

 Variety of communication nodes and "switches" and "gateways"

 workstation

 personal computer

Transmission Links and Speeds (illustrations only):

1 High-speed leased link   1.5–2 million bits per second
2 Satellite   10 transponders at 48 million bps
3 Public data network   9.6 thousand bps
4 Fiber optics short link   200 million bps
5 Intrapremise local area network 2 million bps

Switches:

Multiplexers: share high-speed lines among slower input/output lines
PBX: private branch exchange: telephone switch; advanced ones handle computers as well as phone
Controllers: traffic cops for clusters of workstations
FEP: front-end processor: interfaces a "host" computer to the network
Gateways: advanced switches that handle aspects of routing, accounting, conversion of formats, and network interconnection

many different types of devices, transmissions links of differing speeds and characteristics, and specialized switching equipment they require. The figure does not, however, show the complexity of the software involved in coordinating the flow of transactions across the system.

The biggest single problem that telecommunications managers face is that there is no one network. There are many. They are incompatible. "Incompatibility" means "This dealer will have to buy a special terminal to access our system," or "We have to have duplicated transmission facilities between locations A and B because the two networks use different protocols," or "We have one of everything. An IBM bisync network, one running under SNA, an electronic mail network, several different bus local area networks, an X.25 international network, and . . ."

The history of modern telecommunications is the move from separate facilities and incompatibility toward integration. The logical endpoint, which is already technically feasible though not yet available for everyday operation, will include the following features:

The multiservice workstation that can access any available remote service. This means no terminals "dedicated" to specific computer processors or networks. The workstation may be a "telset," a combination of a telephone and a terminal, which can handle both data and voice. The workstation is the access point for a growing range of electronic transactions, services, and information.

A *worldwide set of transmission facilities* that can deliver all standard types of information—telephone calls, images of documents, memos and messages, computer data and transactions, and pictures to the workstation. Some are leased by the firm ("private" or "virtual private" networks) and others are used on an as-needed basis ("public" data networks).

Switches that make the network and its services transparent to the user. "Transparent" means that the person need have no knowledge that the network exists, just as the telephone system is effectively invisible (if it is working properly). The switches are the "gateways" into the overall service base. They are rather like airport hubs; they coordinate commuters flights, long-haul flights, international regulations, security, etc.

A *set of standards accepted by all manufacturers* of computer and communications equipment and providers of communications services that will eliminate incompatibility. The vendors will not have to standardize their products; photocopying machines vary immensely in cost, size, and function but accept in the United States the standards for paper size, 8-1/2 by 11 inches and 11 by 14. Telecommunications standards define the "interfaces" between facilities, not the facilities themselves.

Moving Toward Integration

Standards are the key to interconnection and the evolution toward integration, but defining them has been and will continue to be an appallingly complex task. There are just too many existing incompatibilities. Countries and leading manufacturers have a vested interest in controlling or owning the standards.

The growing priority for all firms is to escape the current chaos of incompatible facilities. Because so many of the new business opportunities rely on the multiservice workstation, sharing of transmission facilities, and interconnection between networks, their immediate priority is to build their telecommunications architecture around an integrated computing resource. Corporate treasurers, for example, are not willing to use seven different terminals, as they had to in the 1970s, in order to communicate with five banks' cash management systems and two brokerage and foreign exchange systems. And manufacturers and distributors cannot afford to exploit the opportunity of putting a terminal in the dealer's office if that still means several terminals plus different networks for ordering and deliveries, accounting, and sales administration.

Since the architecture *is* the strategy, the business manager's job is to define the business criteria for specifying it. If the integrated services digital network (ISDN) were in place, there would be little problem in doing so. Nor would there be if all vendors had already adopted common standards, if the firm did not have a variety of special purpose systems in place; if the cost charged by AT&T, its new competitors and international Postes Telegraphiques et Telephoniques (PTTs), the national monopolies that regulate telecommunications in each country, were predictable and stable, if all computers used IBM's operating systems software; if, if, if. . . .

As it is, the need for an architectural, not a utility approach becomes clearer with every competitive minute, and compared with 1980, progress toward integration has been substantial. Many of the remaining obstacles are technical and economic. For example, firms have been slow to try out voice/data integration, carrying telephone conversations and computer data through the same switches and transmission facilities because they do not want to put their telephone system at risk or gamble on unproven technology or on vendors who may not survive in the increasingly volatile telecommunications market.

Other problems are more political. Deregulation changed the rules of the game for suppliers and users of communications. Governments across the world see telecommunications as a growing aspect of national economic policy. They are also moving beyond their traditional role as provider of basic telephone services and transmission links. Some of them want to be leaders in markets that build "value-added" services onto those links, services such as videotex, office automation, teleconferencing, electronic data interchange, and electronic banking.

The key problems, though, are managerial. The telecommunications managers' time and their staffs'—and too much of their attention—have to go to managing network operations. For them, nearly every technical component—terminals, switches, and transmission links—poses a problem. Their architectural options are constrained by past investments and old technology. (One major insurance firm's 1987 analysis of its telecommunications needs is entitled "Towards our 34th network.") This is a difficult time for them. They are all trying to achieve some degree of integration and are struggling to make the transition to a new set of technologies.

They are well aware of costs. Telecommunications budgets are growing rapidly in most firms and there is constant pressure from management to keep them under control, which is natural and reasonable but can make it hard to define an architecture and blueprint for the long term. Telecommunications managers recognize that they may not understand the business issues as well as they could and should. They wish their senior managers were a little more appreciative of the technical issues.

Defining the Business Criteria for the Architecture

The architecture has to reflect the business vision. There are a number of often conflicting criteria for selecting the standards, scale and type of technology, and technical policies for the selection of vendors and equipment. The main ones often end with "-ability." Integrability is not the same as integration. Integration is the end point; integrability refers to degree or amount.

Integrability.   Will this blueprint allow us to deliver more and more services to the same workstation through the same telecommunications facilities to exploit business opportunities or improve the way we run our business? The question presupposes a vision of the services, the market, and the firm.

Connectability.   Can customers and units of the firm access the highway system cheaply and simply, without having to add specialized equipment? From what locations—gas stations, shops, office, the home? Again, the manager has to have a guiding theory of the business to decide how important this is.

The standards for connectability are fairly well established. They include the international X.25 standard, which has been adopted by most public data networks; there are, however, many X.25 dialects, just as Boston English is not quite the same as New Orleans English or British English. The variants produce unexpected incompatibilities.

Shareability.   This is currently beyond the capabilities of any multivendor computer and communications system. It involves more than just moving data through switches and delivering them to a single workstation. That is telecommunications-specific and refers to integrability. Shareability relates to the computers that process transactions and manage data.

"Can our sales processing system share data with our manufacturing system?" or "Can we bring together all the services we provide to customers that run on different computers and have different data files and formats?" This is the area where computing and communications have to converge.

Do we build our communications architecture around our computer processing and data systems? Again, what is the business opportunity that would lead us to pay a premium for shareability, the most complex aspect of defining a long-term architecture?

Phaseability.   Can we add capacity, move toward integration, and extend the range of services in increments or must we take a Great Leap Forward, with all the costs and risks that implies? One main reason for designing an architecture is to get away from both incompatibility and from having to make constant changes to existing systems as we expand the highways and the traffic on them. This problem is most acute in firms that have a multiplicity of existing communications networks.

Availability. There are many gaps in the standards, many competing ones, and too much unproven equipment. Many vendors have announced products that may or may not be delivered or that will take years before they are stable. This is as true of IBM and AT&T as of smaller firms.

Managers would like to be able to base an aggressive business strategy on stable technology. Can we do so or must we commit to "paper" operating systems and emerging, but not established, standards? There is no fully proven hardware or software for large-scale computer-integrated manufacturing and electronic document management, for instance, but there are plenty of business opportunities that need it. What assumptions do we make about the ones that are proposed, promised, or available in preliminary versions? When do we commit to them, given that the lead time for implementation of applications that are built on them can be years?

Reliability and Maintainability. Complex systems are complex to design, implement, and operate. They often do not quite work. New ones may work but horrendously inefficiently. The field of computers and communications is full of discrete euphemisms. "Inefficient" really means "so expensive that we can't afford it" or "so slow it can't provide acceptable service and response times," just as "incompatible" means "We can't use this with that even though common sense says we should be able to." The best known euphemism is "95 percent debugged." That is a polite evasion for "It doesn't work at all."

Do we have the technical talent to be able to deal with bugs and new "releases" of software? Are the standards and technology implicit in our architecture widely enough established that we can draw on vendors and outside resources for help when we need it? What are the business costs to us of any problems, even minor ones, in the network operations? The more our business vision centers around direct access by customers to our electronic services, the more important reliability and maintainability become. Security is a subset of reliability.

More and more cautionary tales will be told in the future about what happens when a communications system collapses. In one major developed country, the entire banking system was affected when a software "bug" made it impossible for a top bank to process its transactions. This affected transfers to and from other banks. It was literally bankrupt for a night and the other four largest banks had to lend it about $5 billion. NYNEX, the regional phone company in the North East, had a key switch knocked out of operation in a Long Island hurricane, thus knocking out dozens of businesses. Stories like this will soon be as common as today's "Teenage hackers break into Pentagon/bank/credit card/airline network."

Translating the architecture into detailed technical design, implementation, and operations can be expensive and difficult. That is the job that telecommunications managers and staff are hired to do. Their role is not to answer the business questions that determine the trade-offs among integrability, connectability, shareability, phaseability, and reliability or maintainability. The planning agenda that ends up with the architecture and that reevaluates it periodically is only as useful as the quality of the business inputs. That may be why so many first-rate companies have not exploited telecommunications as well as they could. There are no business inputs.

HOW MUCH DO SENIOR MANAGERS NEED TO KNOW ABOUT THE TECHNOLOGY?

The technology is not a peripheral issue. The planning process has to begin with business strategy but must end up with boxes and cables. If telecommunications utility managers too easily get locked into operational details and sometimes seem unable to move away from a purely technical perspective, at least they make things work.

The opposite extreme, the Great Conceptualizer, can be far more dangerous—proposing alluring and adventurous business ideas but overlooking just how hard it is to translate the grand scheme to reliable operations. Technical people are generally very suspicious of ambitious moves involving telecommunications. Often they are too cautious, but they have reason to be wary of proposals that ignore the issues of lead time, technical practicality, efficiency, and reliability. They are likely to block their ears to the claims and urgings of people who lack technical credibility and experience.

How can senior executives separate business vision from fantasy? How little do they need to know about the technology to make informed judgments?

Telecommunications technology involves relatively simple concepts that quickly expand into complex details. The (necessary) jargon that comes with the description of the myriad components that have to be linked together to provide a service can numb the mind. If, though, the discussion stays at the level of simple ideas, it quickly becomes simplistic, and the relevant details become so divorced from the concepts that the whole picture loses focus. The gap makes it hard for technicians and managers to understand each other.

It makes no sense for managers to try to learn the nuts-and-bolts of the technology. The field is too fast-moving for even specialists to maintain their expertise. What managers need is the same level of basic understanding that anyone involved in business has to have about accounting. They should have a sense of the major categories and terms, analogous to debits, credits, depreciation, and so on. They must be able to interpret a communications plan in the same way that they read a financial statement. That level of knowledge in no way qualifies them as experts but helps them use experts effectively and allows them to take part in discussions of planning and not be scared off because they do not feel qualified to hold an opinion.

The Five Main Components of the Telecommunication Base

However simple or complex a communications system is, it has five building blocks: transmission links, switches, terminals, the network, and the network architecture.

Transmission Links—the pipes that carry the coded signals that contain the information. These are the links along which signals are sent plus the techniques for coding information and making efficient use of available transmission capacity. The medium may be terrestrial cable, microwave radio, satellite, optical fiber—anything that can carry the signal.

Until a few years ago, the cost and speed of transmission were a constraint on large-scale commercial use of telecommunications. New technologies, especially fiber optics and satellites, have changed this and it is no longer the strategic element in network planning. One standard fiber link can carry about 6,000 simultaneous telephone conversations. The comparative costs for "twisted pair" telephone cable is 30 times the equivalent fiber investment per unit of capacity and for coaxial cable (high-speed cable for data communications) the cost is 45 times the fiber cable cost. (The price of transmission may not reflect costs. In the United States, deregulation and new competition have pushed price close to cost, but in other countries, tariffs may not pass on the advantages of new technology to business users.)

Switches—analogous to the airport hubs in the airline systems, for taking off, landing, and connection. Switches are the specialized equipment at "nodes" of a communications network where the signals from transmission links are processed in order to route them to the receiving device (a terminal or computer in the case of data communications and another telephone for voice communications), translate them from one message format ("protocol") to another, improve the efficiency of transmission, or handle other aspects of network operation and control.

The degree of "intelligence" in the switches has become the key strategic element and the most expensive part of the telecommunications investment. They may be thought of as telephone exchanges for computers as well as telephones, with facilities to allow data from low-speed devices, people typing at terminals, to be interleaved on high-speed lines through "multiplexing," and to handle different telecommunications message formats' protocols. The newer switches include software for accounting, error-handling, protocol conversion, and "digitizing" telephone conversations so that they can be transmitted more efficiently.

It is in this area that the technology is moving fastest and is most unstable in terms of products, standards, and vendors. A key question for telecommunication managers is when or if to commit to investing in the switches that integrate voice and data, telephones and terminals.

Terminals—the originators and destinations of the communication. The devices that access the communications facility may be computer terminals, personal computers, telephones, word processors, a variety of workstations, or large, central, "mainframe" computers. Generally, they are personal computers or workstations linked to a remote computer for processing transactions, accessing data, or communicating messages to other terminals. For this reason, the computer's operating system, the software that directs its multiple activities, is a key concern, as is the interrelation between the operating system and the telecommunications architecture.

The Network—the directory of addresses that indicates which terminals can be in direct contact for sending and receiving communications. A "network" is a system of switches and transmission links plus a directory of addresses of terminal points that can access them and thus contact each other. The U.S. telephone system is one such network, with the phone number providing the address. The same telephone can link into other countries' networks even though those systems use different transmission features and conventions; special switches translate the protocols. In the same way, a single terminal may access many networks, and many networks may be interconnected through switches.

The Network Architecture—the set of conventions or "standards" that ensure all the other components are interrelated and can work together. This architecture is the overall design blueprint for creating and evolving the network over changes in time, technology, uses, volumes, and geographic locations.

International organizations are working to set standards rather like the specifications for electrical facilities (voltages, plugs, currents, outlets) that will make it possible for all network equipment and networks to be interconnected. Most major U.S. firms and most computer and communications vendors have adopted IBM's Systems Network Architecture (SNA); most international ones the X.25 standard. SNA and X.25 are key terms in the communications planner's vocabulary. Very roughly, SNA may be thought of as the architecture designed for computer-to-computer and terminal-to-computer communications. X.25 is oriented to terminal-to-terminal communications. In the international field, Open Systems Interconnection (OSI) is a proposed and gradually emerging standard for data communications and Integrated Systems Digital Network (ISDN) a broadly agreed standard for the phone and data utility of the future.

ORGANIZATION AND IMPLEMENTATION: TAKING CHARGE OF CHANGE

What Has to Follow the Planning Stages

Vision, policy, and architecture bring senior managers into the planning loop. Others can then take over the process of design and delivery. What they have to do is challenging:

- Make the case for investments in the telecommunications "highways" that are all cost and no benefit and that may take many years to provide returns. It is the traffic they make possible that creates the payoffs.

- Completely reorganize the technical functions of voice communications, data communications, and data processing, since integrated technologies require integrated thinking and an integrated Information Services organization.

- Establish a set of standards and procedures that balances the need for central coordination without intruding on local autonomy in decision-making and implementations of new systems. The architecture needs a corporate architect.

- Create a pricing mechanism for internal and external users that is predictable and equitable, provides incentives for efficient use, and does not burden initial users of the infrastructure with a disproportionate share of the long-term investment cost.

- Help disseminate the business vision and communicate the technical architecture so that middle-level business and technical staff can work together to get the best of their own and each others' skill in exploiting the telecommunications resource.

- Solve some urgent technical problems in the short term, such as forecasting capacity requirements, managing proliferating and incompatible systems, upgrading existing facilities, and selecting needed new equipment at a time when the products and the vendors are unstable.

- Help staff at all levels—senior executives, middle managers, supervisors, office workers, customer service personnel, secretaries—prepare for an era where more and more of their own work will be mediated by telecommunications: personal computers linked to data and processing systems, workstations, communicating word processors, electronic mail, and more.

All this adds up to radical organizational change.

Organizing for Change

Telecommunications is intrinsically linked to organizational change. As with almost every aspect of computer-related technologies, the organizational issues are likely to be far more difficult to tackle than the technical ones. The scale and pace of change are apparent in the ratio of workstations to people in large firms. In 1987 it was about 1 for every 5 workers in the United States and some Fortune 500 companies were bragging that it was closer to 1 to 2 in their organizations. Leading companies in information-intensive industries like banking and insurance and information-intensive functions in manufacturing firms, such as finance and customer service, have already reached a 1 to 1 ratio.

This represents an immense change in the nature of work. Telecommunications throws the terminal into new cultures. There is no level or function of the organization whose activities are not becoming or have not already been mediated by the workstation. Telecommunications has become a fixture of the office landscape.

Ten years ago, computer people paid little attention to nontechnical organizational and "behavioral" issues. Now, the need to "manage change" has become part of their litany and senior managers are very aware of the complexity of making new systems work organizationally as well as technically.

With telecommunications, change has become the norm. It does not have a beginning, middle, and end. There is hardly time to relax before the next wave of change washes ashore. Almost by definition, when telecommunications is used for business innovation, the intended goal is radical organizational change: one workstation for every three or two workers or even for every one in the near future. The management challenge is to take charge of change, not just manage it. "Manage" means "react to."

In the 1970s computers were mainly confined to the back office and to automating clerical functions. Poorly designed systems, clumsy implementation, or inattention to organizational issues could badly disrupt whole departments, but the impacts, both good and bad, were localized. This is not at all true with telecommunications. It neither automates the status quo nor is confined to the back office. It is a social technology that changes work and the internal processes of entire organizations. The firm's business health will be affected by the competitive implications of telecommunications, and its organizational health by how well it takes charge of the change process and redefines the planning and management process.

Here again, senior managers have to lead, not delegate. They are responsible for the business health of their firm and for its organizational health. Telecommunications today is intimately related to both. It is time to move. The rest of this book is about how to do so.

II SEIZING THE BUSINESS OPPORTUNITY
The Best of Current Practice

Telecommunications opens up new opportunities for business. The problem is how to recognize them.

The experience of firms that have already used communications technology to gain a competitive edge can fuel the formation of an adventurously realistic business vision for telecommunications. Adventure is essential—breaking away from the status quo. Realism is vital—avoiding fantasy and naive technobabble.

Exemplars often also provide early warning signals: what has happened in another industry may very well happen here. Senior managers tend to scan their environment narrowly, focusing on traditional, direct competitors, often overlooking innovations in other industries which might be instructive. A few of the general lessons that might be learned from specific firms' exploitation of telecommunications are as follows:

Financial Services. Electronic delivery can redefine the basic level of service in an industry and shift the business emphasis from marketing separate products to managing the customer relationship through the workstation and credit card.

Manufacturing. Using the telecommunications infrastructure to coordinate production and distribution and linking the dealer to the central organization has reduced key inventories by about 20 percent in many cases. Computer-integrated manufacturing (CIM) rests on

39

having a comprehensive telecommunications capability and a plan for integrating work stations, computers, and data. CIM is a critical survival factor for U.S. and European firms in competing in the global marketplace.

Airlines and Related Travel Services.   Telecommunications is so central to making it easy for the customer to access the firm that any advantage in lead time translates to a long-term technical edge. It has also become a central element for airlines maintaining control over their distribution system. Ownership of a major reservation system has been the battleground in the many 1987 mergers and strategic alliances in the United States and in Europe.

Distribution-Intensive Industries.   Putting a terminal in the customer's office differentiates standard products through electronic services and provides a way of capturing the bulk of the customer's business previously divided among many suppliers.

Petrochemicals and Related Multinationals.   The firms that already have a large-scale telecommunications infrastructure can use it to bypass their banks and control their suppliers. Oil companies have also seized the opportunity of EFT/POS to replace credit cards.

Retailers.   Ownership of the customer contact point—the check-out register—and the communications network is the base for major market innovation and adding services traditionally controlled by other suppliers. This is especially true of financial services now and travel services soon. In addition, electronic data interchange (EDI), substituting telecommunications for paper documents, has already become a basic way of doing business in the grocery industry. Twenty-five percent of all purchase orders, billings, and so forth are now done electronically.

Publishing.   Information can become a premium good if it is made instantly accessible through telecommunications. The lead time for creating a delivery base, though, is long, and being there when demand takes off means moving early. Not knowing whether the demand will occur makes this a high-risk venture. The publishing industry has had many of the most striking successes in using telecommunications to gain an edge and has had a disproportionate number of fiascos.

DEFINING THE COMPETITION

Many of the exemplars raise an additional fundamental and truly strategic issue: Who is the competition? Telecommunications has stimulated major incursions into markets that traditionally are part of another industry: retailing into banking, banking into publishing, airlines into hotels, customers into suppliers' domains. There are three main reasons why this has happened and can be expected to happen more often.

First, telecommunications facilitates the delivery of multiple services at the same workstation and encourages firms to intrude into other industries' traditional territory. That is why so many aspects of retailing, banking, insurance, travel reservations, and publishing are converging.

Second, whenever time and information are key product differentiators or open up new market opportunities, those firms that are ready to deliver electronically will succeed. The strong fill the vacuum left by the weak.

Finally, deregulation breaks down traditional bounds between industries and allows the firm with a strong telecommunications infrastructure to move very quickly into new business areas. A store can become a bank without having to build bricks-and-mortar branches; the point-of-sale workstation is an electronic branch. A bank can become a publisher in the same way without having to open a newsstand; the terminal in the treasurer's office is the purchase point.

The next three chapters provide concrete illustrations of these general points. Each chapter describes one of the three distinct generic strategies that firms have used in exploiting telecommunications:

1. *Get an edge* within the existing marketplace and product base by using telecommunications to redefine the quality and nature of customer service.
2. *Run the business better* than is possible when coordination, communication, and timeliness of key information are constrained by time and geography.

3. *Reposition the firm* by exploiting ownership of a large-scale delivery capability plus the advantage of lead time to change products and move outside the existing market and product boundaries.

These broad strategies overlap with Michael Porter's well-known model of competitive forces (detailed in *Competitive Advantage: Creating and Sustaining Superior Performance*, New York: Free Press, 1985), which identifies five major factors: rivalry among existing competitors, the threat of new entrants, the threat of substitute products or services, the bargaining power of suppliers, and customers' bargaining power. Telecommunications for business strategy is a specific case of a more general competitive dynamic.

Models like Porter's, or related ones that look at using telecommunications and processing to add to the "value-added" chain in the sequence from raw materials through to sales and service, do not answer such questions as

- How to maintain a barrier to imitation when differentiating a product through telecommunications

- How to anticipate which competitors move in from outside the industry

- How to use bargaining power with suppliers through telecommunications.

Exemplars highlight proven application opportunities. They show what has been achieved, and organizing their experiences into general patterns highlights strategic trends across industries that can be anticipated and exploited.

Just as much can be learned, perhaps, from mistakes, the experiences of firms that misread the market, gambled on unproven technologies, failed to unite technical and business personnel, or had the right business idea but the wrong technical strategy. The focus in the next chapters is on opportunities and, almost inevitably, they will often obscure the one hundred fiascos that preceded the one triumph. By definition, strategic innovation cannot be easy. Telecommunications combines heavy capital expenditures, long lead times, and technical uncertainty and volatility.

THREE GENERIC STRATEGIES

There are exemplars in every industry but their strategies and the outcomes are generally not industry-specific. The same patterns emerge in very different contexts. Only the timing varies.

Figure II–1 summarizes the three main patterns and serves as a roadmap for the descriptions of each strategy in the next three chapters.

Look for an Edge in Existing Markets

Put a terminal in the customer's office to provide a direct link to the firm's service and product base. This is often most effective in mature markets. Dealer order entry terminals, electronic cash management and pharmacy terminals are examples. They can be used to provide a two-way flow of services: orders and inquiries from the customer to the supplier and management information from supplier to customer.

Differentiate a standard product through service and information and thus add premium value to a commodity or make it easier to use. Automated teller machines do nothing new yet are seen as an innovation. Airline reservation systems that also handle hotel and car rentals add convenience to standard services.

Improve customer access to the firm by eliminating delays and intermediate steps, especially through direct guaranteed telephone links—no busy signals—or access to information on products, prices, and availability without having to go through a person. Self-reservation systems and special customer service links are examples.

Run the Business Better

Manage distributed inventories by increasing central coordination without intruding on decentralized operations and reduce information "float," the time gap between something happening and being able to get the information about it and respond to it. This allows large firms to combine the benefits of both centralization and decentralization.

Figure II–1.   The Business Opportunities Telecommunications Opens Up.

The Business Opportunity	Key Business Messages
LOOK FOR AN EDGE IN EXISTING MARKETS	
Put a terminal in the customer's office.	• Telecommunications can redefine the base level of service in an entire industry.
	• When a firm gets a good enough system in place ahead of its competitors, it can be hard to displace, even by a better one.
Differentiate a standard product through service and information.	• Service and information byproducts can add premium value to mature products.
Improve customer access to the firm.	• In markets where the product is a perishable good (reservations, foreign exchange trading) a busy signal means a lost sale.
	• Customers often want to contact Customer Service for information, not to complain.
RUN THE BUSINESS BETTER	
Manage distributed inventories.	• Telecommunications allows central coordination of decentralized operations, substantially reducing inventory levels.
	• The savings often pay for most of the cost of building a telecommunications base.
Link field units to head office.	• Field sales and service staff can keep in touch with the office (and vice versa), allowing them to spend more time out in the field.

Figure II–1. continued.

Improve internal communications.

- Without a communications highway infrastructure, there can be no effective strategy for realizing the immense potential of office technology applications.

Improve executive information.

- Most management information systems provide historical data organized around the financial reports cycle. Telecommunications allows managers to get more timely operating information pulled to the center from decentralized business units.

FIND SOURCES OF MARKET INNOVATION

Make preemptive strikes.

- When a successful innovation based on telecommunications changes the competitive structure of an industry, catch-up is measured in years.
- Who is your competition now? In five years?

Create new services by piggybacking and network interconnection.

- The cost of adding new traffic to an existing network is often small.
- The traffic is often some other firm's revenues.
- Whole new industries are being created as telecommunications allows integration of previously separate services.
- Traditional industry boundaries no longer apply.

Link distributed field units to the head office. Insurance agents who are effectively independent entrepreneurs have been enthusiastic adopters of personal computers that can be linked to the company's data and processing base. Traveling salesmen travel more effectively when they can pick up waiting messages and reply to them at a pay phone at the airport.

Improve internal communications in general through all the types of traffic—communicating word processors, document interchange, electronic mail, teleconferencing, and so forth—office technology makes practical. Office technology has been slow to fulfill its promises mainly because the highway has to be in place first. The telecommunications network is the strategic element, not the applications.

Improve executive information by pulling to the center key operating figures from all the business units so that managers can base their decisions on timely business information, instead of on historical financial reports that tell them about the outcomes of particular trends and events when it is too late to affect them.

Find Sources of Market Innovation

Make preemptive strikes that gain immediate customer acceptance so that competitors are forced to respond but lack the technical base to do so; these are the moves that can transform the competitive dynamics of an industry.

Create new services by piggybacking and network interconnection. Add new services at low cost to an existing communications infrastructure and exploit the opportunity to deliver a range of separate services to a single customer contact point.

The Risks and Lead Times

The risks and lead times increase with each approach. Finding a competitive edge involves limited business risk by the very fact that it looks to build on existing products in existing markets. The risks relate to timing. Where telecommunications has already become a business necessity, the risk comes from not moving. The greater the initial investment in the infrastructure, though, the harder it is to move fast, and there is always some degree of risk in wide-scale introduction of technology, however well proven it may be in other firms.

Building a new dealer network is not something that can be done quickly or with guaranteed immediate results.

Running the business better requires a substantial telecommunications base to be in place. It involves a range of risks, many of which are organizational: facilitating changes in the nature of the work and helping nontechnical people adapt to the introduction of unfamiliar and often threatening tools and procedures. Many of the applications that are loosely classified under the misleading term "office automation," for example, involve a mix of computer media, such as voice, unformatted messages, documents, computer data, and pictures.

The telecommunications capacity needed to handle this mix efficiently and effectively can grow faster than it can be installed. Some of the media push the limits of available software to manage them. Complexity, volumes, response times, reliability, and compatibility all compound the strain on the system. Because information highways soon suffer from problems of urban sprawl, there has to be a strong transportation policy and first-rate telecommunications operations for office technology to be a benefit and not a problem.

The third strategy, repositioning the firm, by definition involves high business risk. If it were easy to do, everyone else would be doing it. There lies the return as well. It also takes years. The most reliable estimate is that any major market innovation involving telecommunications that establishes a barrier to imitation is likely to take from five to seven years to complete. Almost all such innovations make it essential for senior managers to ask: Who is our competition now and who may it be soon?

The strategies overlap. The firm that has built a comprehensive telecommunications resource can exploit economies of scale, technology, and expertise and change the dynamics of innovation. It can also alter the economies of innovation. The cost and time needed to add a new type of traffic to the highway system are far, far lower than for the competitor who has to go through the learning curve and investment to build the base. And the same base that it uses to deliver traditional services electronically may carry new ones; an obvious example is the airlines reservation system becoming a hotel and car rental system. The telecommunications architecture is a blueprint for much, sometimes even most, of the firm's business future.

One point that is obvious from the examples of specific industries is the extent to which telecommunications brings them into competition with each other. Perhaps the most interesting questions for senior management are Who will the competitors be in the future? and What do we do when our customer becomes a competitor?

Many of the examples are old ones, by the standards of the telecommunications field at least. They have been chosen as exemplars because the firms gained a distinct business advantage. This means that if the examples are surprising to any senior manager reading them, that person may have quite a lot to worry about; time is passing. The radical ideas of 1982 were the common sense of 1986 and the painful object lessons of 1988. The investments of 1988 are the payoffs of 1993.

3 LOOKING FOR AN EDGE IN EXISTING MARKETS

It is becoming harder to gain a competitive advantage or differentiate one's product in many industries. The market is mature, and prices, manufacturing techonology, and product features are basically the same, and there is little customer loyalty or brand identity. Service has become the factor that makes the difference. In any industry where the customer responds to flexible and fast access to the supplier, tele-communications is the means for redefining service. Three tactics have been used successfully.

1. *Put a terminal in the customer's office* and thus provide a direct link to the firm's service and processing base.
2. *Differentiate the service* by adding information by-products or by making it easier to use.
3. *Improve customer access* to the firm.

The goal is to eliminate time and geography as barriers to service and to use the terminal to redefine the customer relationship.

BENEFITS OF THE TERMINAL IN THE CUSTOMER'S OFFICE

Creating a direct link between supplier and customer has several ben-efits. It allows the supplier to exploit any advantage of occupancy; once in place the workstation should be hard for a competitor to dis-lodge if the quality of service is high. This means it must be reliable, cheap, and able to handle large volumes of transactions without increasing "response time." This is the time interval between pressing the key on the terminal that signals the completed entry of a piece of data or a request and getting an answer back.

If there are bottlenecks in the communications or computer system, response times quickly increase from the typical target of under three seconds to minutes. Sudden increases in volumes over a few months or peak demands at particular times of day easily cause the network to "crash."

It can be a disaster to put the terminal in the customer's office to improve service and then run into technical problems that the customer experiences immediately and directly. If the quality of service is good enough, however, there is no reason for the customer to switch. There may even be a cost in doing so: relearning procedures, for instance.

Transaction costs can be shifted to the customer by having the customer enter and validate transactions; many companies have been able to reduce staff or handle significant increases in volume because the customer replaces their effort or eliminates their paperwork. "Front-end" data capture and electronic transfer also reduce errors and delays; the farther along in the pipeline of paperwork and processing an error is discovered, the more it costs to track and repair. The workstation is the front end of the process and most errors can be trapped at that point.

Using the customer contact point to add services and information by-products further differentiates the firm and locks competitors out. It can serve as a two-way link, delivering reports as well as taking orders.

The customer terminal makes it as simple, direct, and easy as possible for the customer to contact the supplier for orders and to get information, locate stock, and get answers to questions and complaints. The customer can place orders directly and get them processed immediately. This is like having a personal sales force available. It translates fairly directly into reduced inventory levels. Time is money and money is time.

The technical problems involved in doing all this are not trivial. Few of them are strategic, and many relate to the critical importance of efficiency and reliability. The most difficult step is building the initial network and processing base into which the customers are connected. That takes from two to five years typically. Extending it or adding new features takes from one to three years.

Successful Examples

Financial Services.   Electronic cash management systems in international banking are the base for turning a standard service—funds transfers—into a new delivery system and then adding a stream of products. Few of the products are new in themselves but they become so because of the terminal and telecommunications link. Many require little computer processing; they are communications-based and time-dependent. They include reports that monitor foreign exchange, target balance managers, multiaccount balances, and market analyses. In retail banking, ATMs and credit cards open up new cross-selling of services and access to customers.

In insurance, the insurance agent's portable terminal puts the insurance firm's back office in his or her briefcase. Independent agents are many insurance providers' "customers" rather than the consumers they sell policies to. The electronic back office can persuade the agents to favor their product.

Manufacturing.   Dealer systems in the automotive industry provide fairly immediate improvements in such operations as locating parts, placing orders, and processing warranty claims. In almost every instance, dealer inventories are reduced by 15–20 percent. Ford UK has outsold every other car maker in Britain by applying high technology to every part of the design, manufacturing, and sales process. The company's vehicle locator communications system enables a buyer, walking into a Ford dealership, to determine within minutes whether the particular car he wants is available from stock, not just from that dealer, but from anywhere in the country.

Airlines, Travel, and Shipping.   Travel agent airline reservation terminals had such a strong impact on competition that they became the subject of an antitrust complaint and a critical element in the creation of mega-airlines. Carl Icahn's acquisition of TWA; Frank Lorenzo's purchase of Eastern; Northwest buying Republic; Galileo, the strategic alliance among United Airlines, British Airways, KLM, and Swiss Air; the $300 million Amadeus Group's investment to link the reservations and distribution systems of Air France, Iberia of Spain, Lufthansa, and Scandinavian Airlines; American Airlines' dominance of its industry—all of these in 1987 were reflections of the extent to which reservation systems had become a driving force across the industry. In each instance, the competitive move was to gain a reservations capability or to try to displace a rival's.

Reservation and scheduling systems for agents are opening up a similar competitive battle in shipping, with several suppliers of telecommunications networks looking to franchise the systems to shippers. Paperwork that adds close to $500 to the cost of an international trade shipment can be reduced or eliminated by the networks, and they can link the financial aspects of shipping (letters of credit and trade documentation) with the transportation ones.

Distribution.   Bi-directional links between customer and supplier for placing orders and transmitting inventory and order status reports, cost and purchasing analyses, and the like have set new standards for service and become one of the most important factors in selecting a supplier. The benefit to the customer is that the level of customer inventories can be reduced sharply with this improved means of ordering. A pharmacy, for instance, can transform its own efficiency and flexibility of operations through its supplier's direct electronic link into its service base.

Multinationals.   Petrochemicals use their communications capabilities to become leaders in point-of-sale by linking the gas station register to banking networks. Mobil Oil Corporation's debit point-of-sale (POS) program allows card holders to use their ATM cards to purchase products and services at the 1,900 Mobil stations nationwide.

Retailing.   Terminals used by supermarkets to place orders also update prices and print out new shelf labels; one of the leaders in this field targeted "technologically undernourished" stores, for which the terminal provides a step-shift in service and which as a consequence become dependent on the firm. Electronic document interchange is cutting paperwork for orders and deliveries down by days of delay and yard-high "in" trays.

Publishing.   At a time when every major international bank was racing to get the cash management terminal into the treasurer's office, Reuters was already there selling up-to-the-second financial data, at great profit. Since then, the terminal in the advertising agency has become the access vehicle by which companies selling market research data compete for position, replacing standard, historical data with up-to-the-day information for regular and ad hoc analysis.

Several of these examples are worth discussing in more detail because the lessons they provide apply to almost every industry, now or soon.

Capturing the Customer Relationship

Electronic cash management systems were an innovation in the early 1980s. Now they are close to being a commodity. Only a few banks recognized the nature and extent of the opportunity. Many argued that helping the customer to make electronic funds transfers directly from the terminal and track balances through the day was against the bank's own interests. It meant the client would be able to reduce float and minimize the free cash the bank could use at its own convenience. European banks saw cash management as mainly relevant to the U.S. domestic market, whose size covers many time zones. Others felt that all that was needed was SWIFT, the telecommunications system through which the banks, and only the banks, could send the equivalent of a telex authorizing a funds transfer.

This has been a costly case of myopia. The terminal has become central to the corporate client relationship, as the core lending business has dried up and multinationals have become sophisticated in their handling of cash and hence more demanding of their banks. Citibank's expensive and early commitment to electronic banking in the late 1970s and its creation of a global telecommunications network has been translated into a commanding platform for innovation. By getting the terminal into the customer's office early it was able to use cash management as the thin edge of a very big wedge.

It is now irrelevant for the laggards to argue that cash management is not directly profitable. The terminal is the access and delivery point for the fee-based products that are the revenue generators of the future; these, like funds transfers, are often standard services increasingly differentiated by electronic delivery. In the global banking environment, the move away from lending has meant a move into the financial services products business. Providing global transaction services is now a major focus that requires banks to bring the technological know-how they use to automate their own operations right into the clients' offices. The drive is on to extend cash management toward the integrated treasury workstation, to add automated letters of credit, trade documentation, trade financing, collections, broker-aging, and dealing functions and to sell electronic information.

"To compete and thrive during the remainder of this decade and throughout the 1990s, banks must integrate the various elements of electronic banking," a report from McKinsey and Company and Salomon Brothers noted in 1987. The study predicts systems technology spending at U.S. commercial banks will more than double, rising from $8.2 billion in 1985 to $17 billion by 1990. That would be a far cry from 1981, when banks spent $0.2 billion on hardware, systems, applications software, and telecommunications equipment (*Information Week*, June 8, 1987).

Recent Extensions of Electronic Cash Management

The cash management business, previously confined to large companies and financial institutions, is moving into the middle market (companies with sales between $25 million and $500 million) and advances in electronic banking are accelerating the pace.

In June 1987 Bankers Trust unveiled its Micro Cash Connector, which enables treasurers to initiate payments via a PC. Mellon Bank introduced a funds transfer system in June 1987 that lets corporate clients with a microcomputer make payments over the automated clearing house network at much lower costs than wire transfer.

Manufacturers Hanover's Interplex Business Manager combines cash management with a midsized company's accounting systems. "Several packages do part of this process. But ours is the first that ties it all together in one lump," says Jim Witkins, senior vice president at the bank's financial services group (*Computers in Banking*, July 1987).

Chase Manhattan has expanded on its global cash reporter Info-cash to build the Chase Customer Access Project (CCAP), an enhanced architecture for the bank's global system. With it Chase is aiming to attract companies throughout the world: "If you can reach out with a device and a network into a customer's office and really give the customer, in the convenience of his office, access to the full range of the Chase Manhattan Bank product line, and give him a commitment, there is really no reason why we should not become the bank of concentration for that customer" (*Computers in Banking*, December 1986).

The degree of differentiation created among the international banks between 1983 and 1988 was astonishing. A lot of banks were simply out of the game. The laggards had to race to catch up to Citibank, even though its network was primitive in technical terms, the quality of its software was often well below the professional standards expected of a large firm, and it spent far more money than it really needed to get its systems in place. Citibank's success came from the quality of its business thinking and from its using its lead time well. It will be hard to displace its terminal from customers' offices.

But Citibank has had to work just as hard to maintain its early advantages. On occasion, the poor technical quality of its systems undid much of the business strengths. In one key market, it ran out of capacity and the system was overloaded at peak times. Even though the average availability was over 95 percent, customers could successfully access it between 10:00 A.M. and 1:00 P.M. only three days a week. They saw this as 60 percent availability.

Telecommunications involves both business strategy and boxes and cables. Each is critical. In this instance, without a telecommunications strategy, business strategy is constrained. There are only two European banks that can compete with the top five United States banks, and the shakeout in international securities and investment banking following London's "Big Bang"—a total deregulation—has shown that no network means no business.

The Terminal as a New Theory of Business

Two of the most striking illustrations of why senior managers in any industry have to think ahead and ask what impact the terminal in the customer's office can have on the basic dynamics of competition come from the medical distribution field. One is American Hospital Supply Corporation (AHS) and the other, described in the next section, is McKesson. They contain similar messages.

Both of these stories are old and overexposed. Anyone who has attended any conference on using information technology for competitive advantage will groan at hearing yet again about AHS. But the lessons from its dazzling success in stealing an industry are only emerging now, 10 years later. The story of how AHS took over leadership in its field is an old one. It introduced its Automated System for Analytical Purchasing (ASAP) in 1978 and by 1980 had transformed its industry.

Hospitals that are linked to the company's computer are far more likely to buy supplies from American. And the average hospital order of the ASAP system averages 5.8 items, which compares with an industry average of just 1.7 items per order and with an average of 2.4 items on a conventional order received by American. Revenues show the same trend, with customers spending as much as three times more than they did before, using American's manual system (*Business Week*, September 8, 1980).

Customers reported significant savings in inventory levels; examples were cutting stock on hand from $700,000 to $250,000, and turning inventory over monthly instead of eight times a year. AHS's competitors were put at a clear disadvantage.

In fact, AHS had not really innovated at all. It had applied technology to one of its core business drivers: making sales calls. That is basically all ASAP originally did, in a business where the more calls you make the more sales you get.

An ATM, in the same way, was not really a business innovation. ATM protects the basic core of a bank: getting consumers' daily financial transactions. Airline reservation systems, widely heralded as a radical innovation, are really a conservative evolution; they are an order-entry system that handles the basic business of an airline, which is responding to customers' requests to allocate them a seat on a plane.

This is not to belittle AHS or American Airlines but to highlight a general lesson from 10 years' competition in the electronic marketplace. The most effective wins often do little "new." They focus the use of technology on the core aspects of the customer relationship and build on them. In a way, the winners are the conservatives.

They can, though, be very radical in building on the base they create. AHS continued to do so and exploit the competitive gap opened up through ASAP, by extending the link into its own suppliers' locations. This meant it was able to cut its own inventories by reducing the delays across the whole distribution system that translate into buffer stocks. It was also able to add information services, offer volume discounts to groups of customers, and get better terms from its own suppliers.

Information as a Service.   AHS provided hospitals with management information on inventories, usage of products, costs, and more. This is data that AHS has to collect anyway in order to process the customers' orders. For very little extra cost, it turned the data from an internal part of its overhead to an economic good of value to customers. If a hospital places all its business with AHS, it gets a comprehensive management information system; the Justice Department's antitrust unit found this as interesting as did hospitals.

The use of available operational data to create a valuable new product service is another general lesson in 1988 from AHS. It is discussed in more detail in Chapter 5, Market Innovation, and anticipates the generic strategy of many airlines, credit card firms, banks, and retailers: begin with telecommunications to capture the customer delivery base and then use customer data to create new services. Data about customers created by service transactions is a major competitive resource. American Express and Diners Club are exploiting this, so too is Sears.

Discounts to Customers.   Several hospitals in the same city can combine their orders and get volume discounts. They need not get together to do this; all that is involved is AHS assigning them a common account number used by each of them to initiate an electronic transaction.

Better Terms from Its Own Suppliers.   AHS streamlined its ordering and delivery procedures. It made no sense to have multiple suppliers of the same item if there were no difference in price or quality. AHS put pressure on particular suppliers to reduce prices and in return shifted most, if not all, its business to them. It is able to monitor suppliers' production schedules and inventory levels and check them against AHS's projected needs, through its own computers and communication links to their computers. If a supplier's projected stocks are inadequate, AHS alerts the firm to the situation. The supplier takes notice very quickly.

The main question arising from AHS's experience is Will the competitive edge provided by information technology really be sustainable? In the early 1980s AHS came under strong pressure as the fat margins hospital suppliers were used to were eroded by medical insurers and hospitals demanding tight cost containment. AHS was bought by Baxter Travenol, after an unsuccessful proposal to merge with Hospital Corporation of America.

Every infohype fad generates a counterfad. On the conference circuit it is now as fashionable to belittle the AHS story as it was in the mid-1980s to talk it up. Some commentators argue that much of AHS's success was luck, in terms of timing and customer response, and that it was not really planned but just happened.

Yes, but. . . . AHS created immense competitive disadvantage for just about every other firm in its industry. Even Johnson and Johnson, well-managed and with huge resources, was still trying in 1984 to find a response. When Baxter Travenol acquired AHS, its CEO stated that having AHS's electronic delivery base would have added $400 million in sales to Baxter. Obviously, no single business advantage can be sustained forever, and it is important to recognize that environmental shifts can remove old advantages: in a context of cost containment by hospitals, sales calls are no longer such a core business driver as they were before the new strategic issue became control of expenses. Lead time gave an edge for years. Time changed some of the environmental factors that helped provide that edge. In any case, a new message from this old story is that even if the leader's competitive advantage may only be for seven years, the other's disadvantage is for seven years plus, where "plus" may be another seven years.

Links from Customer to the Firm's Own Supplier

Foremost McKesson, a distributor to pharmacies, made similar moves, with similar results, only a few years later. McKesson had been a lackluster firm, and it took a change of leadership to stimulate the conscious search, not for new sources of business, but for better performance within the traditional one. McKesson has almost completely eliminated "information float" and "hand-offs" between the customer and itself, and between McKesson and its own suppliers.

It guaranteed delivery within a specified time if the customer placed the order via the terminal; this shifted costs out to the customer who, of course, was able to reduce inventories because of the reduced and guaranteed turnaround time.

Its own computers handle McKesson's ordering. In 1987 McKesson planned to spend $115 million on updating its computer and building efficient new "hub" distribution centers reported to be bigger than 12 football fields in size. McKesson's 17,000 customers use a simple hand-held calculator-like terminal to place their own orders. The computer they access ships the request to one of the distribution centers. The central data base contains records on over 2,000 manufacturers' delivery timetables, order quantity, and related information. Fifty percent of McKesson's own orders are placed by the computer's calculating the company's requirements as orders flow in from the over 17,000 locations and sending the order electronically. McKesson, McKesson's own suppliers, and McKesson's customers can all balance their inventories with their sales.

McKesson also took over, for a 5 percent commission, the problem of collecting payments from third parties, mostly insurance firms, on behalf of its customers, electronically, speeding up collection time. It is an electronic middleman and the largest third-party processor of claims for drug payments in the United States. This is an example of the new economies of innovation opened up by having the terminals and network already in place.

American Hospital Supply Corporation and McKesson together illustrate many of Michael Porter's competitive forces:

- *Rivalry among existing suppliers* led to the two firms using this new vehicle to create an edge.

- *The bargaining power of customers* put their competitors into a weak position.

- *The bargaining power of suppliers* has been used by both the companies, who have also exploited their strength in relation to customers to put immense pressure on their own suppliers.

- *The threat of new entrants and threat of substitute products* means that the firms continue to face challenge. AHS has standardized the customer ordering process and the electronic equivalent of the paperwork. There are a number of firms that can attack AHS by offering electronic services that adopt those standards so that the customer has no switching costs. These include Proctor & Gamble, IBM, GEISCO (General Electric's subsidiary that has become one of the most aggressive players in the electronic market), and Reynolds and Reynolds, which sells computer systems to hospitals.

Customer-Initiated Transactions: The Cost of Doing Business

The customer service terminal changes both parties' cost of doing business. One firm calculated that having the customer input and validate the data ships out about 35 percent of the transaction cost. McKesson was able to reduce its own purchasing staff from 140 in 1976 to 13 in 1981 as a result of putting terminals in 32 of its main suppliers' offices.

One of the main benefits to the supplier of having the terminal in the customer's office is cost avoidance, rather than cost displacement: the ability to handle higher volumes without increasing staff. With manual order processing, service units generally have high incremental costs. As volumes grow, costs increase proportionately, with limited economies of scale; there are often even diseconomies of administration and coordination. The main component of costs are salaries and related overheads.

When a firm invests in the communications and processing infrastructure to change the nature of service, it substitutes technology, which is mainly a fixed cost, for labor costs, which are variable. This means that, within given ranges, transaction costs are volume-independent. If the firm is a high-volume operator, this gives it an edge. Once it is above the breakeven point, the advantages of fixed-cost technology over incremental labor translate into profits. If, as in the case of AHS, the system in itself draws in business and increases volumes, the profits are compounded (see Figure 3–1).

Where service costs are a large portion of the operating base, the terminal in the customer's office gives a large edge to high-volume operators. They can become the low-cost producer through their technology base, especially if they can exploit the many available economies of scale and efficiency in telecommunications.

Of course, the initial investment can be high, or it can take several years before the payoffs are apparent. Moreover, cost avoidance is hard to measure, because costs actually increase simply because volumes are increasing, but at a slower rate than they would do otherwise. One bank, which was the first in its large national market to introduce ATMs, gained occupancy and has held it. The direct revenues from its ATMs are close to $50 million. Much of this is from charges paid by other banks. Their customers use the leader's ATMs more than they use theirs, because of more and more conveniently located machines. More importantly, however, the bank would have had to add between $100 and $150 million to its cost base to service its current volumes without the ATMs, or forgo the business. This example, where the costs avoided are larger than the direct added profits, is typical—and compounds the problem of anticipating and measuring business value. At one level the $100–150 million is not "real" savings. Or is it?

The cost benefits from electronic transactions can be substantial. One automobile manufacturer has reduced the service costs for a car purchase and delivery by over $200 through terminals in the dealer's office for use in handling purchase orders, invoices, handling warranty claims, financing, and other details.

In corporate banking, a few U.S. banks are quoting fees of $6 where the average charge for the transaction is $16–20 for the rest. SWIFT makes a profit on a worldwide transaction at its guaranteed price of 40 cents. This is about 15 cents more than the average local phone call and shows the cost dynamics of the electronic banks versus the manual processors. A bank has to send only one telex or make two international phone calls to be at a disadvantage of not just 50 percent but 5,000 percent.

The Message from the Terminal in the Office

Whether viewed in terms of revenues or costs, the terminal in the customer's office raises important business questions, with timing being a key aspect of each one. Is responsive, fast, direct service likely to become a major differentiator in our industry? When? What is the lead time for building the delivery base? What is the impact of moving too early versus too late? To move early we have to be sure of customers' response, know that the technology is stable and affordable, and be able to get the volumes needed to cover the additional fixed costs. To be able to afford to wait, we must be sure that there will be no advantage of occupancy and that we at least have the telecommunications blueprint so that we can respond quickly and not have to start from scratch, or worse, think from scratch. Is the American Hospital Supply story just a special case?

PRODUCT DIFFERENTIATION

Creating product differentiation is always a goal of marketing managers everywhere, and telecommunications is hardly the only vehicle for doing so. It has the special feature, however, of creating a barrier to imitation because of the inherent lead time and technical difficulties.

Telecommunications is a valuable differentiator in business where convenience is dependent on access and information. The general strategy can be effected in several ways:

1. *Provide walk-in or walk-up service* as does the automated teller machine.

Figure 3–1. The Impact of Electronic Delivery Base on Costs of Service.

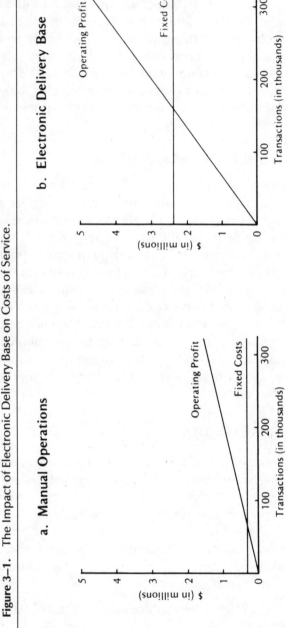

a. Manual Operations

Transactions (in thousands)

$ (in millions)

Operating Profit

Fixed Costs

- Costs are largely incremental
- Low sensitivity of operating profit to changes in volumes
- As volumes increase, must add people

b. Electronic Delivery Base

Transactions (in thousands)

$ (in millions)

Operating Profit

Fixed Costs

- Telecommunications and processing base largely fixed cost
- Profits highly sensitive to changes in volumes
- Can add volumes without increasing people: cost avoidance

Figure 3–1. continued.

This example assumes:

- Average direct profit per transaction (funds transfer, dealer order, securities trade, reservation, etc.) = $20
- Labor needed to process transaction in manual environment = $15
- Operating profit = $5
- Fixed costs = $300,000
- Electronic delivery base adds $2 million to fixed costs; labor reduced to $5
- Operating profit = $15

Ownership of electronic delivery:

- Gives edge to high-volume operators
- Requires heavy capital investment
- Increases business risk: small changes in volumes can have large impacts on profit
- Accelerates the competitive impact when the electronic delivery base creates new business

2. *Add to a standard transaction a follow-on service* that is associated with it but that generally requires a separate activity: applying for a mortgage after signing a purchase contract for a house or checking in at the airport after checking out of your hotel on a business trip.

3. *Exploit the difference between available and accessible data.* Available data requires effort to locate or obtain. There is plenty of it, stuck away in filing cabinets, on computer tapes, in reports. In fact there is a glut of it, a species of information pollution, such as the proliferation of computer-produced reports that move from the data center to the managers' wastepaper baskets in nanoseconds (billionths of a second, a basic measure of the speed of computers). Making useful available data accessible makes it an economic good of value.

Examples of Successes

Financial Services.   The automated teller machine does nothing new. It cashes checks, reports account balances, and accepts deposits, and yet it is seen by the customer as an innovation. This is the example of walk-up convenience.

First Bank of Boston's Shelternet and TRW's Loan Link are instances of adding a follow-on service. They are used by real estate firms to apply for loans on behalf of their own customers. They are linked into credit files, so that a loan commitment can be made within 30 minutes. Letters needed for credit verification are mailed out within two hours. The entire process is done in 2 days instead of 30 to 45. Again, the systems do nothing new, but they add convenience and save time and effort.

Manufacturing.   One heavy machine manufacturer provides a link from the customer's equipment to its central computer. When a machine fails, a signal is automatically sent and diagnostic information is returned immediately. Spare parts are dispatched automatically as well. The manufacturer's field service unit is alerted. This is a simple and imaginative business idea that in itself does not sell machines but adds a small increment of convenience that has tipped the balance in a number of competitive bids.

Airlines, Freight, and Trucking.   Scandinavian Airlines owns the Hotel Scandinavia. When customers check out, they can get their boarding pass and check their baggage. The check-in counter is in effect in the hotel's terminal at the front desk. The Scandinavia's occupancy rate has increased substantially.

The hotel is the same as before, as is the price. The new service brings together two parts of business travel that are separate but associated, annoying time-wasters. Saving time and preplanning is the most common contribution telecommunications can add to such standard services; it makes trivial transactions welcomely trivial to get done.

In the freight and trucking industries, convenience is not a trivial matter. The CEO of Pacific Intermountain Express (PIE) recently commented that "In trucking today, we all use the same highways and freight terminals. Our only competitive advantage is to stand out technologically." PIE developed a system that lets customers access its computer files to check the status of shipments. This is data that PIE has to maintain for its own operations. It was available to customers by phoning PIE's customer service unit. Making it accessible turns it into a marketing asset.

Distribution. American Hospital Supply Corporation used data as a differentiator in the same way. In any business where the cycle of ordering, scheduling, shipping, and forecasting is a core determinant of profits, it is now common sense for suppliers to stop wasting data. By selectively opening up their computer files, they put a customer service representative in the terminal.

Petrochemicals. Electronic funds transfer at point of sale (EFT/POS) has for years been recognized by banks and retailers as one of the major business opportunities and challenges of the coming decade; yet, it has been the oil companies who have moved quickest, not because they see this as a major market innovation, but because it is a useful differentiator for both the customer and the gas station dealer.

There are three key aspects to EFT/POS.

- *Debit instead of credit payments.* The customer's bank account is immediately debited and the seller thus has the funds available at once instead of having to wait. In many countries credit cards account for 30 percent of gas station sales. Mobil was able to give its customers 4 cents per gallon discount as part of its Mobil + system piloted in 1983 as CashFlow and extended to 5,000 gas stations in 15 states in 1988.

- *The card as the base* for providing the service. Mobil spent $30 million on building its Mobil+ point-of-sale network that connects with banks' automated teller systems' communication switches. When customers use their bank ATM cards to purchase Mobil gas, their accounts are debited after the transactions have been validated electronically.

- *The card as provider* of additional customer privileges. The Mobil+ system allows cardholders to withdraw "instant cash" from ATM machines at the station, and it also offers a free VISA card and a toll-free travel reservation service with lodging, air travel, and car rental rebates.

The advantages of EFT/POS at the gas pump are so large, with benefits to dealers, customers, and the oil firm, that even though progress has been sluggish in banking and retailing in this area in many countries, in the oil industry it has become a race. In Belgium, New Zealand, Norway, and Brazil (where EFT/POS has substantially reduced robberies), gas stations are being automated at as rapid a rate as ATMs were installed a few years ago.

Publishing. The difference between available and accessible information translates into major market opportunities in this industry. Dun and Bradstreet is able to sell the Official Airline Guide on-line for twice what it charges for the printed one. The Chemical Abstracts Service's experience gives another measure of the premium. The rate of renewal of subscriptions to its journal has dropped by 50 percent; people prefer electronic access even though it costs much more.

The information must be useful, of course. The most successful applications of telecommunications in electronic publishing have been those where the data is of value in itself and telecommunications adds convenience, makes it easier to obtain information where needed, or where timeliness is crucial. Efforts to persuade people to subscribe to electronic newspapers or catalogs have not met with success however. A printed newspaper is marvelously convenient, portable, and cheap. Why change?

Telecommunications as a Natural Differentiator

Many other instances could easily be given here of communications-intensive customer services that significantly improve the firm's business options for what was previously or was likely to become a commodity. Adding more examples, though, would be overkill; the mind blurs with yet another example of what too easily sounds like a "so what?" issue—"So, United Pomegranate lets kiwi fruit growers in the Southeast get access to its database of supermarket deliveries. So what?"

By definition, product differentiation is not striking. If it were, it would be product innovation. There are some important—and on the whole rather disturbing—patterns running through the decade of firms' successes and failures in using telecommunication for product service and differentiation. They seem to be basically the same in terms of the technology and one would expect that the steep declines in the cost of software would have speeded up the diffusion and success rate.

After all, it is now absolutely obvious that telecommunications is a natural differentiator:

1. Convenience and walk-up service are always attractive features to customers. Who really wants to make a special trip to the bank just to get money to spend at the supermarket?
2. Customers are increasingly responsive to service providers who can cut out delays and cumbersome procedures; travel agents or consumers do not prefer to phone the airline then the hotel and then the car rental company.
3. Every firm in a fast-moving business needs fast feedback on what is happening in its marketplace and wants to track time-sensitive operations. If it can get easy access, simply and quickly, it will always be ready to give business to a supplier who provides that data.

All this means that every large firm in nearly every industry clearly has to factor telecommunications into its strategic plans for its core products and services. As markets mature, as new entrants bring down margins and prices, and as customers are offered a broader range of choices, a comprehensive telecommunications strategy is a competitive necessity, not a luxury. If a firm does not protect its ability to differentiate, someone else will use the opportunity.

The disturbing pattern in the evolution of the electronic marketplace is that in 1988, despite numerous examples, most firms seem to be loping, not running, to the future. Inertia still seems to prevent aggressive action. Perhaps senior management awareness of the competitive opportunity has increased, but not the willingness to turn commitment into action. The need to invest in telecommunications is seen as important but not urgent.

Part of the reason for this is precisely because product differentiation is not striking and hence often not seen as "strategic." Investing in business innovation via information technology has long lead times—five to seven years, not one—and there are too many short-term pressures on middle managers for them to lead the process of change and no reasons for them to follow when their own senior executives are not leading.

Telecommunications is no longer an issue of technology but of taking charge of change, instead of just reacting to it. That is very hard to do without leadership. It is even harder when a leader is blocked. The following disguised case is at least as much a moral fable as American Hospital Supply, a decade later.

The insurance industry has been a laggard in exploiting telecommunications and runs the risk of losing control of its market; regulation is its main protection and if that is eroded, banks, retailers, and even airlines are actively waiting to add insurance to their other services delivered via the workstation. Insurance firms' profits from employee benefit insurance—group medical, accident, and life—have already been severely eroded by large companies' sophisticated efforts to get the best deal possible and manage their costs. Alternate suppliers have intruded on the traditional providers. The entire insurance industry has been through years of an underwriting crisis.

In that context, product differentiation is obviously vital and, given the lessons from banks with ATMs and cash management, and from airline reservation systems, it is just as obvious that telecommunications is both the base for differentiation and essential for retaining customer loyalty.

Jack Allen is the vice president for marketing systems in Alpha Mutual (AM), one of the top 10 U.S. insurance companies. He has worked for AM for 20 years, mainly in business planning functions. He was given a new job by Tim Parnell, the director of marketing. Douglas Baker, the director of information systems, was concerned that no one in AM's information systems department seemed to be looking at long-run strategic opportunities and decided to build a capability within Marketing. Baker gave Allen a fairly small budget, $300,000, out of which he was to develop a customer workstation, largely in response to a rumored pilot project being run by another major insurance firm. He had only a year of funding, so he used his imagination and off-the-shelf hardware and software, a commercial time-sharing service, and "dial-up" phone lines.

The system, BENEFIT, is for use by benefit managers in customer firms. It has many features, including the ability to review which 5 percent of employees pulled down the highest medical claims (using data AM has to collect and store anyway as part of its own operations), a range of claims and eligibility information, benefit managers news, and software the manager can use to create his or her own applications on the personal computer.

Allen spent under $200,000 to develop BENEFIT, which sells for an annual fee of $10,000. It is new and well ahead of competitors' products, which are either focused on clerical operations, not the benefit manager's needs, or cannot handle large volumes of data, or do not provide free-form access to data. Initial customer trials were very favorable and in late 1986 BENEFIT was in use in four customer locations.

Allen's funding has been cut off. The system is ready to go to market, but there is no champion within AM. Baker, his sponsor, has left the company. Marketing systems has been moved out of the marketing department into information services, where Allen's new boss is unsupportive, and Allen feels he is now seen as an enemy by his old friends in marketing as "I am now an information systems (IS) guy. I tell them I am a hostage, but they remember the story about the first thing the Arab did after leaving the tent—he turned around and pissed back in.... AM didn't get to be number eight by being a leader. The first thing they want to know is who else is doing this? If no one else is, we aren't going to do it."

Allen feels AM still has a 12 to 24 month advantage—12 months if competitors develop the system his team did and 24 if they use traditional information systems development methods. He feels that customers are not price sensitive. "They only shop around when you start giving them lousy service—then they become price sensitive. The system is a great competitive opportunity to hold existing customers and sell new ones." He is keeping the product alive and hoping for support somewhere.

This example seems more typical than not. To it can be added the example of the leading oil company whose marketing managers acknowledge the need for a point-of-sale strategy but are doing nothing to make it happen, though they will help it happen or at least let it happen. They cite cost pressures and senior management conservatism as barriers to action. Similarly, an aggressive head of information systems in a major U.S. manufacturing firm gets lots of interested responses to presentations on the opportunity of full-scale dealer automation, but no funds. Etc., etc. Reading and talking about information technology and competitive advantage seems in 1988 to be like discussions on the federal deficit or nuclear disarmament: "Someone should be doing something about this."

Picking the Winners: The Customer Decides

The only valid justification—and admittedly a strong one—for not doing something is the history of failures that has to be acknowledged along with successes. The critical factor in the success of telecommunications innovations is customer acceptance. There is no point in differentiating a product in ways that leave the customer indifferent. All the firm has done is add to its cost base; telecommunications is not cheap and organizing data and storing it online is the most expensive and difficult aspect of data processing.

This point ought to be obvious, but there are dozens of examples of well-designed, well-implemented innovations via telecommunications that quickly and completely fail because the expected customer demand is just not there. The failures in telecommunications look very much like the successes: Home banking is still a solution searching for the problem, while automated teller machines have been enthusiastically adopted everywhere. The *New York Times'* effort to create an on-line information service was as disastrous as Dow Jones' and Dun and Bradstreet's ventures have been successful.

In 1982, Citibank and McGraw-Hill, two consistent winners in the electronic marketplace, launched an automated trading system for energy products. In 1984, they abandoned the venture. Merrill Lynch and IBM similarly in 1987 wrote off the $240 million they had invested in INMET, a telecommunications network for security dealers. Federal Express followed its successes in using mobile communications with Zapmail, a state-of-the-art network for sending facsimile via satellite communications. It was a flop.

All of these ventures were sound in concept. They were launched by firms who had previously been very successful in the ways described in this and later chapters. The differences in their good and bad experiences obviously relate to the customer's response, not to the technology. Too many management teams—in business units and IS—are far too far away from their customers. Bank executives rarely see the customer, or their front-line customer service people who work face-to-face with them. Airline marketing managers are discovering they do not know who their customers are. Oil companies see the retail customer as an abstraction. Insurance firms who sell three main lines of personal service—life, home, and car insurance—have little idea of which customers buy only one of these. The ones who have started to build the on-line data bases to find out are easily able to cross-sell the others.

All these firms do market research and claim to be "customer-driven," but they too often do little customer research. They also tend to think in terms of what customers ought to want and while huddled in the head office come up with a stream of brilliant innovations and products in electronics. Steven Mott, an executive of MCI, now a consultant and a shrewd observer of what went wrong and why in the first generation of these products, examines a number of gaps between expectation and outcome in Electronic Information (EI) services (this summary is from a forthcoming book by Mott and this author).

Electronic mail has become a $1 billion dollar industry but customer demand is way below forecast. Telex was expected to die quickly, telephone tag become a thing of the past, and huge volumes of letters be siphoned away from the postal system. Only about 20 percent of telex has shifted to E-mail.

Mott summarizes the lessons here:

First, existing products and services are almost always satisfying and fulfilling certain important needs. Those who would propose advances in technology must identify, acknowledge, and fill those needs if they would ever hope to supplant loyalty to such established communications rituals as telex and use of postal systems. Second, work habits are tremendously difficult to change and must not be trivialized or minimized in the enthusiasm to "show off" new technology. Third, it takes a lot of understanding and a lot of retraining of users, particularly those who are new to the process altogether, to gain adherence to a new way of doing things, especially one as important as communications. Fourth, efficiencies—whether cost or functional—must be real. It won't take users long to realize something is not better, faster, or cheaper. And finally, market evolution comes in stages, with different customer targets, product orientation and positioning, and provision requirements at each distinct stage. E-mail should have been developed for the large communications user first.

On *home banking* Mott says:

Banks failed to see that the functionality they were providing was inadequate compared to the hassle and time it took to do some of the functions electronically. Ditto for bill paying, as many users found it was easier and quicker to do it manually. Home banking hasn't proved to be convenient; it is not faster, better, or cheaper yet, so long as local branch banking with working ATMs is easily available.

On *at-home investing* (portfolio management via a personal computer linked to a central computer and data bank), Mott observes:

What many of these EI vendors continued to forget was that the conceptual sell was easy, while the implementation proved daunting. It was almost automatic getting investors attracted to the idea of investing by computer—especially at home; creating a service package that was easy to understand, fast and simple to operate, and complete with the necessary functionality was the way to make money.

Just as with changing old habits in messaging, getting people to use a complicated computer program/service to replace what they generally did over the phone or by reading had to have the right amount of perceived value for the cost and effort. And just because computers are available doesn't mean they will work for all products. High numbers are very seductive; real numbers usually are sobering. Nor does the presence of a computer mean the presence of a knowledgeable user. (Statistics continue to show that the average PC is used less than one hour per day for less than one-tenth of its designed functionality.)

The real business lesson from this experience, though, was the failure of EI vendors to recognize the relatively distinct segmentation of customers' and buyers' behavior. The computer revolution was happening, but in stages, with relatively small pockets of skilled and well-equipped users. Investor behavior was also segmented, with a premium segment that was extremely self-sufficient which became saturated with at-home investment products very quickly. EI dreamers were lulled by the numbers and the possibilities of the technology but forgot the value they needed to create and the reality of waiting for fundamental change in behavior.

Among *trading systems* was the Citibank/McGraw-Hill much-trumpeted GEMCO. Comments Mott:

GEMCO's parents found that changing a worldwide culture was enormously difficult. Overcoming normal computer phobia was bad enough; getting vastly different participants to trust an electronic system for trades would have been a missionary development. Volatile markets make poor testbeds for new technology, especially if it is expensive and difficult to understand and learn. And once again, sheer numbers of transactions do not translate directly into capturable, sustainable business. Particularly not in a market that can double or halve virtually overnight.

Mott has this to say on *broker workstations*:

What the two behemoths—IBM and Merrill Lynch—missed (besides not acknowledging the difficulties of working together) was the need to provide substantial functionality with demonstrable ease of use. In other words, the new system had to be obviously better than the collection of price feeds, printed materials, and databases brokers currently used. And the system needed to be implementable without substantial disruption or training requirements to the broker's actual function.

Brokers trying out IMNET found it reduced productive selling time available rather than boosting it. Finally—and perhaps most poignantly—the decision to implement such a system on IBM hardware and software technology by definition limited any specialized application system's immediate capabilities and downstream expansion potential. IBM technology in the current era helps businesses maintain the status quo; rarely is it flexible and powerful enough to change the status quo in a competitive-position enhancing way.

Kiosks, such as workstations providing mutual funds, IRA and insurance transactions, and information at airports, are similarly seductive in concept:

Again, the dependency on evolutionary technology based on conventional programming proved deficient in delivering a good marketing idea to an impatient, demanding set of users. The attractiveness of the concept proved so seductive that vendors and providers rushed out with the systems before studying how, why, and where potential users might want to work with them. And the lure of huge usage (at reduced marketing costs) made vendors less sensitive to the costs involved in making such a substantial change in business delivery systems.

The single and central message from contrasting the successes and the failures is a simple and resonant syllogism.

- Customers control the electronic marketplace in determining the impact of service and product differentiation via telecommunications.

- Telecommunications is a natural differentiator.

- Therefore, the winners will be the firms who talk to customers, not about them, who understand the difference between customer research and market research, who use telecommunications, workstations, and operational computer transaction processing systems to collect data about customer profiles, buying patterns, demographics, and trends, who ensure that the people closest to the customer—the salesforce, customer service supervisors, etc.— play an active and initiating role not a reactive one in product and service development. The best ideas for exploiting telecommunications as a differentiator often come from the field, not the head office.

Many firms will have to rethink how they handle product planning when telecommunications becomes a potential source of differentiation and innovation. Who comes up with the ideas? How can the firm assess likely customer acceptance of a service that does not yet exist? Pilots for home banking suggest there is a market; the demand does not occur when the full system is operational. By contrast, almost every bank underestimated how easily and quickly people would respond to ATMs. What is the basis for making forecasts when one is inventing the future?

Business managers who have risen from the ranks of the marketing department are likely to look for traditional sources of differentiation and thus overlook opportunities, or they underestimate the technical costs and difficulties of implementing a telecommunications-dependent system. Technical staff may have plenty of new ideas but not understand the characteristics of customers, distribution channels, and marketing, which will decide whether the idea is just a gimmick or a source of real competitive edge.

The need to close the knowledge gap between the business and technical worlds is most acute when the opportunities telecommunications open up relate more to product differentiation than to market innovation. Innovative large-scale initiatives get close senior management scrutiny. They may still turn out to be dumb ideas, but at least the best talent in the firm pays attention and provides inputs. Using telecommunications for differentiation of standard services is at a humbler level of attention. It is very easy for middle-level marketing managers to be totally unaware of opportunities waiting to be picked up.

It is just as easy for technical staff to drive the bandwagon that management jumps on once it sees telecommunications as important to the business future, without really understanding the business issues. That surely explains why so many pilots for home banking and electronic publishing look like wishful thinking and show a poor understanding of the customer's needs, expectations, and attitudes.

How can firms get ideas from the market back to the technicians and vice versa? How much do the firm's marketing managers know about what is happening in its own and other industries in terms of telecommunications subtly changing the concept of service and allowing firms to steal market share unobtrusively? How do they find out?

Some answers to these questions are given in Chapter 10. They involve both education to mobilize for change and, above all, systematic cross-fertilization between the technical and business parts of the firm. The need is for the hybrids who are fluent about both business and telecommunications technology. Those who are fluent in both areas are either adding a digit to their present salaries as the head-hunters move in or looking for venture capital.

CUSTOMER ACCESS

Both the terminal in the office and differentiation through walk-up service improve the customer's ease of access to services. There is a third facet of using telecommunications to get an edge within existing markets that more directly focuses on access to the firm as a strategic issue. This is particularly important in markets where the product is time-dependent. An obvious example is foreign exchange trading, in which the market changes literally in seconds. There is more variation in prices across short time spans than across the whole market at any one point in time. In such instances, quality of communications is key. If the customer gets a busy signal or the trader has to say, "I'll call you back," a deal is gone. The size of the firm's market depends on the ease with which the customer can contact the firm. Telecommunications obviously is a general vehicle for improving access.

In the instances cited in this chapter, someone in the firm took a new look at old operations and saw telecommunications as a business issue, not a technical one. They are then likely to find gold in unmined territory. Examples are:

Customer Service Contact Numbers.   Toll-free telephone numbers encourage the customer to call and have changed the nature of customer service units. For example, Procter & Gamble receives 900,000 letters and phone calls a year about its products. These are mainly requests for information, complaints, or testimonials. A special service unit codifies the data, providing P&G with a direct early warning signal that alerts it to problems it might otherwise not have found out about for months.

Special Access Links.   The newer computerized telephone switches (private branch exchanges, PBX) and data communications links are being used to provide key customers with an access line that virtually never results in a busy signal. There are several examples discussed below where this has been the base for getting substantial increases in market share.

Telephones and Profits

A striking example of the strategic impact of improving customer access comes from a British securities firm, "Hachman Brothers" (a pseudonym). HB was under immense business pressure in the mid-1980s. The London market, which had been heavily protected, was due to be opened up to outside competition. Trading in securities was moving away from face-to-face dealing and off the floor of the exchange. There was a flurry of mergers and acquisitions and the entry of well-capitalized foreign firms, who were also much more efficient in their operations than the leading British ones, including HB.

HB saw the urgent need to improve its position and looked at the traditional ways of doing so: recruiting top specialists from other firms, acquiring smaller companies, and beefing up its back-room operations. All of these options were expensive and in the end unattractive.

HB hit on a new source of growth—its own phone system. Much of its client base is outside Europe. It analyzed its clients and segmented them by current profitability to the firm and the potential profitability if communications were improved and trading volumes increased as a direct result. It grouped them into the four categories shown in Figure 3–2.

Previously, all customers had the same chance to get through to HB's dealers. If those in quadrant C of the diagram, low current and potential profits, tied up the phone lines at peak times, those in quadrant B, the prime clients, were unable to contact the dealers and went elsewhere. This happened too often.

HB could not afford to give instantaneous and guaranteed direct access links to every client. Instead, it installed a PBX that

- Gave priority to incoming calls to the dealers on the basis of their point of origin, which meant that at busy times, the prime customers were never put on hold

- Provided a few key clients with "dedicated" lines, reserved for their use only, which meant that they never got a busy signal

- Eliminated the need for operators

Figure 3–2.   Using Telecommunications to Provide Differentiated Levels of Customer Access.

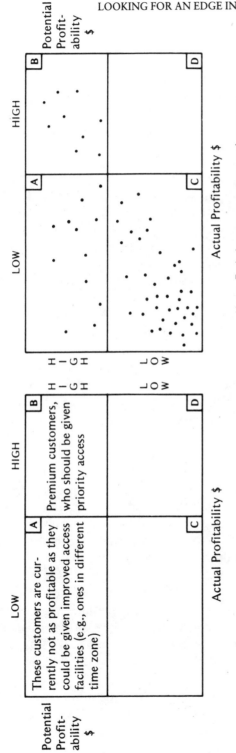

Note: Each dot represents an actual customer

- Potential profitability is estimated profit if this customer has immediate and guaranteed access to services
- If customers in quadrant C tie up the phone lines, business from those in B may be lost
- Customers in A are likely to increase their business with the supplier if they are provided with a terminal or direct phone links
- Customers in B can be given priority service through telecommunications
- The cost of providing telecommunications access can be high (e.g., for a car manufacturer to put all its dealers on line): need to set criteria and priorities for selecting customers

- Allowed the dealers to select among different incoming calls to pick up the prime customers first
- Routed calls to the dealer automatically even if he or she was not at home (much of HB's business came from South Africa and the Far East, across many time zones) or at another desk
- Improved every aspect of response time and virtually eliminated busy signals.

The idea here was to match the quality and ease of access to the category of client. Many companies who have put terminals in the dealer's office or who provide telephone reservation services give them all the same chance to tie up the system. HB's simple matrix is a useful way of identifying where it is worth investing in the (very) expensive technology that segments the access links.

HB's senior managers had simply never thought of its telephones as a strategic business resource—nor had its telecommunications staff. It took an outside consultant, Keith Bellamy, to raise the idea. Bellamy's own management thought this was a silly idea. The PBX cost more than HB's main computer and represented the largest single investment it had ever made in information technology. Telephones were considered part of a firm's operations and overhead, not a source of market opportunity. For HB, however, telephones and profits go together. The investment had paid off well and gave HB a strong competitive boost.

Several years later, in late 1986, the London securities market was completely deregulated, transforming in a few months every aspect of international financial markets. Very few U.K. firms, including the bank that acquired HB, realized how critical telecommunications and computing would be in managing what was a discontinuity, not a change. In August 1987 the *Economist* reported that the backlog of unsettled trades was $10 billion versus less than $2 billion a year earlier.

One leading player, in a Top-Security-Burn-Before-Reading memo sent to its top managers (and naturally leaked), admits that it cannot tell what its risk position in the foreign exchange and securities markets is within eight months. It trades a sizable fraction of its capital daily in the global market thanks to the wonders of telecommunications.

It had invested in dazzle technology for traders but had not paid any attention to the real issue: if telecommunications is used to speed up trading and taking on risk, it must be used to speed up settling and managing risk. Every U.K. player in the city of London saw the importance of using telecommunications for trading. Almost none had a clue about the need to look at the whole front office to back office to settlement process: As the *Economist* reported: "Already many foreign investors prefer wherever possible to buy and sell British equities in New York, where there are few delays." They use the wonders of telecommunications to do so. HB was perhaps more lucky than meritorious in its earlier success. The credit is owed to the young consultant; management took the credit but did not learn the broader lesson.

Gaining Market Share through Ease of Access

In the extremely competitive British holiday tour market, bankruptcy is quite frequent. Tour operators buy hotel rooms and air flights in bulk and sell them at very low margins of profit, often at short notice when it rains in Britain and shines in Costa Brava. Thomson Holidays works with over 3,000 travel agents, who used to phone in to one of its 16 regional centers to make reservations. The local phone exchange was often overloaded, and Thomson lost business as the customer or travel agent lost patience. Even though the problem was caused by the phone company, it was Thomson whose image for service suffered.

Putting all the travel agents on-line into the computer reservation system was too expensive to be practical. Instead, Thomson installed a videotex system that they could access via a cheap terminal through a special phone line, using a local call and bypassing the regional offices. Videotex is a simple way of organizing information in "frames" so that it can be easily retrieved and displayed on a television screen; each frame is in effect a page. The system can handle 20 messages per second, with an average response time of 3 seconds.

Thomson's reservations staff were no longer swamped with queries that also tied up incoming phone lines. The agents could get information easily and quickly, in far less time in fact than the reservation staff could listen to a question, get the answer, and read it back. Later, Thomson linked the videotex system to its reservation system. Now, most of its travel agent transactions do not involve any human intermediary. It has been able to save 200 staff in the centers. It has also become the market leader in a business where differentiation is very hard to achieve. The videotex system was the differentiator, as better service must be in a commodity business or one that is heavily time dependent. The Thomson story is the European equivalent of the American Hospital Supply story—stealing an industry via telecommunications.

Thomson has also sustained its edge well. In 1987, in a vicious price war with its major competitor that squeezed the industry's profit margins to under 3 percent of gross revenue, Thomson's profits grew by 26 percent. In that year it sold 4 million tours, about 70 percent of which included plane travel on Air Britannia, another Thomson Company, which carried more passengers than KLM, with a load capacity of 90 percent versus KLM's 65 percent.

The key aspect of Thomson's success, though, is that any firm in the industry could have launched the initiative. Thomson used the U.K. Post Office's PRESTEL videotex system at a cost of well under $500,000. Whereas AHS, in the late 1970s, had to push the technical state-of-the-art (and made several expensive failures en route), Thomson could buy the technology off the shelf. It did; the others did not. Every commentary on the Thomson success highlights the role of Colin Palmer, a senior business manager. The technology was available to all; the business imagination was not.

Customer Service Access Links

"Customer service" is often the euphemism for the complaints department. A growing number of firms are recognizing that it can be turned into a marketing tool through telecommunications. The customer who is encouraged to phone in provides the company with a lot of information, and if the call involves a problem, resolving it quickly leads to free advertising.

In 1980 Buick received an average of 80 complaints a day and took 25-30 days to resolve them. In 1984 it introduced a toll-free customer assistance center, which has reduced the turnaround time to 10 days. The local dealer gets a response that it can pass on to the customer in less than a day. This had helped Buick raise its customer satisfaction quotient substantially.

Procter & Gamble's customer service operation was mentioned earlier in this chapter. Because customers usually have the item next to them when they phone, P&G has been able to trace problems with defective products at once, instead of finding out weeks later when the damage has spread. Sony similarly used calls asking how to connect a TV set to a home computer to redesign the product in months. Firms like Whirlpool and GE have saved time, both their own and their customers', by diagnosing problems over the phone, instead of dispatching a service technician.

General Electric found that only 10 percent of calls were complaints. Twenty-five percent were from people wanting information when they were thinking about making a purchase. Its Answer Center receives over 2 million calls a year; this means that GE is able to talk directly to 500,000 potential purchasers.

Every customer call provides demographic data or information on the effectiveness of advertising, which is the stimulus for many of the inquiries, pinpoints specific problems that may be pointers to widespread ones, and stimulates ideas for new products. Many companies store the data in computer files for analysis. Making it easy for the customer to contact them is the source of valuable, up-to-date customer research information.

Of course, all this comes at a high cost. The telephone lines and trained customer service staff are very expensive, especially if the systems are to be made available to suit the customer's convenience, which means outside normal working hours, and are to be continuously staffed. Is it worth it? It depends on the extent to which ease of access translates into enhanced and appreciated service and in turn into sales.

A study commissioned by the White House in 1979 concluded that fewer than 1 in 6 dissatisfied customers complain to firms. They complain, though, to about 11 other people. Satisfied ones talk to 3 others. GE and American Express have no doubts about the return on money spent in this way. Service is the differentiator. The three elements of service are

- *Access*—making sure the customer can get to you

- *Time*—responding quickly

- *Quality*—having a good product in the first place, at a good price.

When a firm's competitors are roughly equal to it in terms of quality, then the other two factors move up the list in importance to the customer and to the company.

REDEFINING SERVICE

Three themes run through all the examples of how telecommunications has been used to get an edge within existing markets.

1. Telecommunications redefines the meaning of "service."
2. The edge comes from business imagination, not technology.
3. The customer decides success and failure in the electronic marketplace.

It is hard to find industries where telecommunications is not already or will not soon be an important aspect of competitive positioning simply because service is always a differentiator, except in the special cases where one product dominates the market. In health care, an anticancer drug hardly needs service to sell itself, but generic drugs do.

4 RUNNING THE BUSINESS BETTER

CENTRALIZATION-WITH-DECENTRALIZATION

Many of the benefits of telecommunications technology come from applying it internally. Firms with an integrated perspective on information technologies have been able to exploit the multiservice nature of telecommunications to improve business functions and save money. The applications cover nearly every activity involving paper, communication, analysis, storing and accessing information, and making decisions that require coordination throughout the organization.

The telecommunications infrastructure is the base for exploiting office technology, end-user computing, and other computer-based vehicles of competition. Too many firms focus on the individual vehicles and have not planned ahead for the communications highway system. They lose many of the benefits of the vehicles themselves because of this.

Four categories of applications stand out as ones that change whole aspects of organizational activity, rather than just making useful but marginal improvements on the status quo.

1. *Managing distributed inventories* by combining central coordination and decentralized operations
2. *Linking distributed field units to the main office* to improve the sales staff's productivity and get information to and from them easily and quickly
3. *Improving internal communications* and exploiting office technology as effectively as is practical when standards are not yet fully established

4. *Improving executive information* by pulling key operating data to the center from remote business locations.

These applications all reveal how telecommunications eliminates the dichotomy between centralization and decentralization.

In managing inventories, the central planning unit can get a complete update on the status of orders and stocks out in the field, without intruding on the autonomy of the field staff. At the same time, the local units can use the terminal, workstation, or "distributed" computer to improve their own operations and get information to and from the center, without delay or red tape.

By linking field units to the main office, the sales staff can keep in touch with the branch or head office and be kept informed of price changes, stock levels, news, and so on, without having to go into the office. They can spend more time selling. At the same time, the central unit can coordinate their overall activities better.

Improving internal communications has the greatest potential impact on the basic aspects of doing business. Substituting the workstation for mail, and even in many cases for telephones, reduces delays. What previously took a week takes a day, what took a day is done inside the hour, and what required an hour is reduced to a minute or so.

One consequence of the barriers of time and distance on organizational information flows is that so-called management information systems (MIS) often provide senior managers with little of the information they really need to run the firm. The reports such systems provide are too limited, too late, and too narrow in focus. They generally report historical financial figures rather than up-to-date operating indicators that can help managers anticipate potential problems instead of finding out about their consequences after the event.

Telecommunications allows instantaneous reporting from the decentralized operating units. The converse is also true, but deciding whether to give decentralized decision makers data from the center raises some complex political and ethical issues for senior managers. It must be recognized that telecommunications may increase central control through ownership of data and the ability to monitor decentralized units.

Highways and Traffic: The Need for a
Telecommunications Architecture

In many firms, expectations from the investment in information technology have not been met. It is as if there is some missing ingredient. Some firms can point to a magnificent soufflé, while others who followed the same recipe end up with a deflated disappointment. The missing ingredient, more often than not, is recognizing the importance of the communications architecture. Almost every aspect of the use of information technology to help increase organizational productivity depends on the interaction between computers, data, and communications.

Telecommunications is often mistakenly seen as a tactical add-on to the strategic design of the application. The visible hardware and software dominate attention. The invisible network is taken for granted. An obvious analogy is the telephone system. Installing a new phone can be done quickly and easily. As far as the user is concerned, the receiver is the key element for purchase and use; the conversations the users create are traffic on the telephone highway system.

It is because that system is already in place and has a master blueprint—the "architecture"—that new phones can be connected to it, lines extended, and facilities upgraded. The job of the telephone supplier is to design, implement, and operate it. Its users give it little thought. They are interested in the traffic, not the details of the highway. But the highway system had better be there in the first place. Every aspect of the economics, logistics, and lead times of telephones changes dramatically if it is not.

Using telecommunications to run the business better rests on putting in place a highway system that

- Links the firm's main business locations

- Is based on a set of standards and policies on selection of vendors and equipment; these are needed to avoid incompatibility of computers, terminals, and network protocols

- Avoids proliferation of communications links and specialized switches

- Has the flexibility and capacity to allow new equipment to be connected, new applications added, and additional volumes of traffic carried efficiently and cost-effectively.

These capabilities are embodied in the telecommunications architecture, a blueprint for the long-term infrastructure. No ad hoc, piecemeal, reactive approach to telecommunications can create an architecture. Firms that have grown their communications systems haphazardly are limited in the range of computer traffic they can exploit to run the business better—until they make it a priority to define and implement the architecture. That requires senior management policy decisions. The longer they wait, the greater the opportunity cost and the more difficult it is to rationalize the growing incompatibilities.

Stand-alone Systems and the Multiservice Workstation

One reason most large firms have overlooked the need for an architectural approach to telecommunications is that applications of computer systems are planned and implemented in isolation from each other. Special-purpose networks are used for transaction processing, electronic mail, or teleconferencing. In many applications, this "stand-alone" perspective is encouraged because there is no immediate need for telecommunications.

The user of stand-alone technology has a desk-top device, which is directly under his or her own control, with independent and self-contained data files and software. A "local area network" may link up personal computers within the same building so that they can share printers and local data stores. This improves cost-efficiency but makes relatively little difference to the range of applications and hence the range of organizational benefits that can be added.

Two very common applications of stand-alone computers are for spreadsheet analysis and word processing. Both have the advantage of saving time and improving productivity. The benefits are limited, however. Adding telecommunications shifts the focus from personal to organizational productivity.

The manager whose desktop computer is linked to a telecommunications system is able to access remote databases, send messages to other personal computers, not just ones in the same building, and obtain and transmit reports. Such a system makes it practical to consolidate departmental and divisional budgets and distribute them quickly, gain access to various sources of budget data, obtain data for analysis directly from the remote storage, and simplify and speed up coordination among various locations.

The communicating word processor, in the same way, handles organizational aspects of managing documents. Typing is only a small part of the birth-to-death costs of paper. One Exxon study found that the average document in the firm's head office was copied 40 times. Of these, 15 were filed away in permanent storage.

Adding communications to a stand-alone system changes the type and scale of the benefits. Those benefits are biggest if the architecture allows the word processor and the personal computer to link to each other and share data and processes, becoming multiservice workstations. Without a network architecture, the best that firms can expect from their investment in information technology will be direct, useful, but limited minor improvements in personal productivity and gains in efficiency of operations from traditional extensions of traditional data processing.

MANAGING DISTRIBUTED INVENTORIES

Typical Payoffs

Most large firms have to carry buffer stocks of inventory because of "information float," the time gap between something occurring out in the field and the information being available to the central coordinating and planning units. Telecommunications cuts that float dramatically, with the result that firms can reduce their levels of inventory by 15–20 percent on average.

The impact of centralization-with-decentralization on managing distributed inventories is shown by the example of Booker Dean, a leading (and pseudonymous) oil supply firm. Booker operates autonomous units in 36 countries. In 1983 its long-term debt was $400 million and rising. Booker's management wanted to improve the quality of money management across the firm.

Cash is just another inventory. Booker's was expensive to carry: interest charges on long- and short-term debt, bank charges, and lost opportunities to profit from investing spare cash for short periods, including overnight. Booker's management wanted to increase local responsibility for managing money and for making sure that funds were collected and disbursed quickly. The divisions should deposit checks on a Friday, for instance, and not wait until Monday.

Decentralization of cash management seemed to be the answer. In an international firm like Booker, however, there has to be substantial centralization to handle foreign exchange trading and hedging, risk exposure, short-term investment, and netting. At any point in time, one unit might need, say, $2 million in U.S. dollars, while another has a surplus of $1.5 million. Some units would need to draw down on lines of credit, but others have spare funds that could be transferred to them to avoid having to do so.

Booker set up separate bank accounts in each country and gave the divisions responsibility and incentives to tighten up their management of their own financial operations. At the same time, the corporate treasury unit ran the 36 accounts as if they were a single dollar account in New York, using a major bank's international cash management system. (To do this, Booker shifted most of the accounts away from other banks who could not provide such a facility.)

Within a year, Booker had cut its long-term debt by 40 percent, a reduction of $160 million. Its senior management ascribed all the savings to the new system. That is a measure of the cost of information float to the firm.

There are many other such examples, in manufacturing and distribution particularly. Toyota of America installed minicomputers in its larger dealers, linking them to corporate headquarters. Being able to get figures on daily sales and stock levels makes Toyota's central staff much more flexible in planning production and distribution and in adjusting their short-term forecasts.

The automobile manufacturing industry has moved fast to exploit the opportunities of telecommunications to slash inventories. Examples include British Leyland's Stocklocator system, which allows a dealer to locate a given make and model of car anywhere in the country in under 30 seconds. Dealers have been able to reduce stocks substantially.

Buick's Electronic Product Information Center (EPIC) system provides dealers with facilities to locate a car, compare Buick models with 132 other makes, configure a car, and analyze lease versus buy options. Pat Harrison, Buick's manager of EPIC, calculates that the 400 dealers who use it (out of 2,950) sell an extra two to three cars a month (worth $144 million a year to Buick). EPIC also significantly reduces errors and inconsistencies in the selling prices.

The figure of a 15–20 percent improvement in productivity occurs again and again in reports of information technology: improvements in productivity from office technology, reductions in head count of corporate staff, reductions in inventories, and cuts in manufacturing cost. Some firms have been able to exceed the 20 percent level: Hewlett Packard reduced work-in-process inventories in its small printers division by 82 percent.

The 15–20 percent figure is useful in establishing the general target of opportunity. So many firms, especially in the automotive business, have been able to reduce inventory levels by this amount that it has to be stated bluntly that any manufacturing firm that has not moved to use telecommunications in this way is guilty of wasteful neglect. Given the 1.5:1 or 1.8:1 inventory-to-sales ratio for U.S. manufacturers between 1959 and 1984, even a 5 percent reduction is a major business contribution.

Managing Suppliers

Very large firms can use telecommunications to control their suppliers. One major aerospace company required its suppliers to install equipment for computer-aided design and manufacturing, linked to its own CAD/CAM system. It can send detailed parts specifications and changes to them electronically. Previously, suppliers' ability to handle rush orders and custom orders distinguished them. By requiring them to use the same design and ordering system and to meet the same quality standards, the aerospace firm has pushed them into competing almost entirely on cost. American Hospital Supply's computers automatically check suppliers' prices and allocate business to the one whose quote is lowest. It also monitors their production schedules and inventories and notifies them if they are inadequate to meet AHS's needs. AHS provided the software to the suppliers that significantly reduces their bargaining strength in dealing with the firm.

Xerox's copier division exchanges quality control data with its suppliers electronically and sends its own manufacturing schedules to them. The suppliers ship parts to meet that timetable, reducing Xerox's own inventories. This has contributed to an estimated reduction in production and inventory costs of 18 percent.

These are specific examples of a larger issue: Who controls which part of the overall business system? When GM announced that it would not deal with any supplier in the medium-term future that had to send it paper documents instead of electronic ones, the entire market moved to adopt GM's manufacturing automation protocol (MAP) standard for computers and communications. When retailers and banks link their networks to provide EFT/POS, is this banking or retailing? Who controls the outlets and gets customers' loyalty?

LINKING FIELD UNITS TO THE MAIN OFFICE

The combination of portable personal computers and telecommunications opens up what is among the highest-payoff investments a firm can make in its efforts to improve productivity: providing a two-way link between field staff, especially salespeople, and their branch or head office. Sales representatives usually have to come into the office more than they would like, because that is where much of the information they need is located, especially the paperwork. When they travel, it can be hard for them to keep in touch with their home base and up to date on price changes, product announcements, and availability of goods. The same is true for field staff.

Historically, computer power and access to computer data have been committed to particular business functions on the basis of physical proximity to the computer center rather than to productivity opportunities. The initial applications typically address accounting, followed by sales, finance, and personnel—all mainly head office functions. When personal computers began their rabbitlike procreation in the early 1980s, the initial applications focused on accounting and finance, and the main users were professional staff. The reasons for this are fairly obvious. As long as computers were expensive, they were applied to large-scale processing applications.

Telecommunications allows firms to process information where the productivity opportunities are—often out in the field.

The personal computer becomes a traveling office via telecommunications and reduces traveling to the office. The statement that this is among the highest-payoff investments a firm can make is not a the-office-of-the-future-is-now assertion. There are innumerable examples to back it up. Hewlett Packard (HP) is a recent one, which both summarizes the benefits and is an outstanding model to guide other firms in how to make sure the technology is sensibly rolled out to the field and made to work organizationally as well as technically. It also gives answers in a positive direction to the question: "If the payoff is so high and the opportunity is so obvious, how come most companies have not yet done anything?" The blunt summary is: "because it needs management attention and effort."

HP's Sales Productivity System (SPS) is built on HP's admirable lap computer, the Portable Plus, linked to HP's information network. The impetus for SPS came from Jim Arthus, vice president of U.S. field operations. In early 1985 he asked Ben Menold, field productivity manager, to look at what could be done to make substantial improvements to sales force productivity. After years of 20–30 percent revenue growth, sales and profits were under pressure. It cost HP $200,000 a year to put a salesperson in the field. Menold took an idea from a senior HP executive, who had used statistical quality control in his R&D and manufacturing assignments and applied the techniques of total quality control (TQC) to come up with an operational metric of "productivity" and baseline measure of work levels that would be used to guide the use and assess the impact of SPS.

The team Menold brought together to brainstorm ideas rejected traditional revenue-based measures, such as market penetration, selling costs per unit of volume, or performance versus quota as incomplete and often misleading in that the results of improved selling would often not show up in the figures for six to eight months. They concluded that hours per customer contact was the best criterion. Sales reps should be out selling.

Baseline data was collected in the four areas (out of 17) where 170 sales reps piloted SPS in the first half of 1986. Half used the system and half did not, to provide a control group. The features of SPS are shown in Figure 4–1. Of the total system cost, up to late 1987, of around $5 million, only $100,000 was spent on software. SPS mainly packages existing Portable Plus software.

The time-measurement process was repeated after two and then five months of use. Sales representatives were given a special watch that beeped every 43 minutes, each representative could then note what he or she was doing and where. Far from being an intrusion or an irritation to clients, the beeper watch drew attention to the fact that HP was working hard to improve productivity. This interested clients, who also saw the sales reps using technology as they tried to sell technology.

The results of the pilot were striking and were confirmed as the full system was rolled out. (In the summer of 1987, 1,500 of HP's representatives were using SPS.) Total customer contact went from 26 percent to 33 percent of the work week, an increase of 27 percent; face-to-face contact stayed the same, at 8 percent. Travel and training time dropped by 39 percent, from 13 percent of the work week to 8 percent. There were other impacts, too. The quotes below (taken from telephone interviews and from several 1987 articles) capture the results:

> Before getting the Portable . . . after a couple of sales calls, each of which would produce half a dozen small action items, the rep would return to the office to take care of these administrative tasks—sometimes as early as 3:00 P.M. Now they can use the Portable to write messages informing people of the order changes, requesting others to call customers, or even write up their call reports wherever they are. With the Portable, reps can feel comfortable scheduling more sales calls, and know they can still get the administrative work done.

> With the portable I can file a memo when the subject is fresh in my mind and then dial into the network from the nearest phone, thus cutting down on office trips.

> The program automatically rolls over to the next day or week the three to five things-to-do that don't get done in a typical day. This eliminated the 15 minutes I devoted at the end of each day revising my paper calendar. If I ask a support specialist for information a customer wants on a certain date, Time Manager automatically reminds me at deadline time.

> It lets me work wherever I am, and that could mean in my home or office, at a customer location, on a transatlantic flight to another HP site in England, or accompanying a customer to a high-level meeting at HP headquarters in California.

> With the type of accounts I handle, I don't necessarily find I'm making a lot more sales calls—maybe only one a week—but the quality of preparation has really improved. It's given me the time to thoroughly understand my customers' problems and really tailor my presentations.

There were no technical problems in implementing SPS, because the highway system, HP's Information Network, was in place, and the lap computers are proven and backed up with plenty of packaged software. There were, though, problems with the traditional guardians of the technology. The informations systems people were "a resistance point" for some of the resources needed for SPS to succeed. The sales representatives needed access to marketing and sales data; some of the IS managers viewed this as "sensitive" and had "built a security shield" around it. Much of this was in fact public data anyway (e.g., pricing and product announcements), but the shield made it hard to get.

There are some useful general recommendations to make about effective implementation from the HP example.

Figure 4–1. Giving the Sales Rep Access to Vital Information through Remote Communication Resources.

PORTABLE COMPUTER

- Personal files
- Address files
- Expense and travel files
- Calendar and time manager
- Spreadsheets

Regional HP3000s

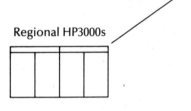

CUSTOMER DATA

- Prospect lists
- Customer product-use information
- History of HP/customer relationship
- Territory information
- Forecasting

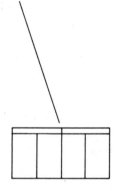

INDUSTRY INFORMATION CENTER

- Marketing suggestions/ideas
- Competitor information
- Related accounts
- Software interface information

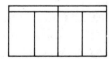

PURCHASE ORDER INFORMATION

- Communication with headquarters
- Price and inventory information
- Purchase options
- Overall sales performance
- New-product announcements

1. Be sure to get an operational definition and measure of "productivity." Many of the disappointments from what *Fortune* in 1986 called the "puny payoff" from office technology reflect the failure to do this. Without it being defined at the start, there is no basis for defining the features and functions of the system, the sales reps are unlikely to buy in to the venture, and no one will be reliably able to demonstrate its resulting business value.
2. Be systematic in piloting and do not rely on the ad hoc prototyping now fashionable with PCs. Know your goals, get out in the field and think in terms of the sales rep who is the real user, not the head office sponsor. Take an entirely field-centered view of implementation.
3. Be systematic in tracking benefits and impacts.

Many other instances of the benefits of linking the field to the branch or head office could be added here. While the specific details differ among examples, the careful planning and systematic piloting and evaluation characteristic of the HP example consistently appear in the companies that are successful in such applications.

This is simply a "must-do" opportunity. The main barriers to exploiting it are lack of leadership, lack of infrastructures, and unwillingness to commit time for action. At HP senior business champions were pushing for a business initiative and demanding justification in business terms. The relatively low cost of SPS resulted largely because HP's Information Network had already been evolved to provide a reliable and comprehensive information utility and delivery base. The elapsed time from conception to operational definition of the criterion for "productivity" to pilot to full implementation at HP was well over two years.

Many are the bright ideas that drift and never find a champion, and thousands of "pilots" get no follow-up action, and thousands of launches move fast then slow then stop. HP's coherent, sustained, and systematic—and very sensible and tough-minded—process means that, as *Sales and Marketing Management* stated in its February 1987 article, "this hi-tech firm will score a beat this summer when it becomes the first major corporation to automate its entire sales force."

A New Perspective

The availability of telecommunications—telephones with data communications—shifts the perspective on computers from the central data center to the remote user. Firms can think in terms of who needs access to services and information, what device can provide it, and what links have to be put in place. With the present strategy, using telecommunications to run a business better, it is the sales force or the field maintenance unit that needs better access to information. In both instances, the technology saves time and money.

The general question raised by the examples here is not "Should sales representatives have personal computers?" but "Which people in the field would be more effective if they were kept more up to date and better informed?" Linking distributed field units to the center should be on almost every large firm's "must do" list. The payoff can be rapid. It is worth the firm's investing about 60 percent of the user's salary if the personal computer, which is the decentralized part of the combination, adds just one hour of effective work a day—reducing paperwork, managing one's calendar, or maintaining customer data, for example. This figure is based on a Honeywell study in 1982 that asked what expenditure would be justified if the equipment helped an employee produce 10 percent more work. It assumed that salaries rise at 10 percent a year, overhead is 35 percent, the equipment is written off over five years, and the cost of capital is 12 percent. The study concludes that the justified investment for a manager whose salary is $35,000 would be $19,600. That buys a lot of computing power. The cost of communications and data plus the central computing resource that handles the access requests from the field machine are much higher. A $5,000 stand-alone personal computer (fully configured, with plenty of software, storage, and a high-quality printer) needs an extra investment of around $15,000 to turn it into a multiservice workstation. Communications adds about 30 percent to the basic cost; the rest is central storage, processing, and staff support.

IMPROVING INTERNAL COMMUNICATION

The most interesting question about the Office of the Future is why has it been so delayed when it has been clear to every innovative manager that word processing, electronic mail, facsimile, elimination of paper, etc., must improve overall efficiency. Few firms seem to have been able to put all the pieces together and diffuse office technology across the organization at the rate they expected in the early 1980s.

The gap between expectation and outcome results from the long lead time between investment and payoff, lack of a clear vision, and lack of a telecommunications architecture. How to hurdle these obstacles is the topic of the balance of this chapter.

Keeping in Touch: Rethinking the Status Quo

Advancing beyond limited pilots and localized applications is a problem in office technology. It is very easy in a pilot system to get immediate benefits, which if extrapolated to cover the whole organization would add up to a considerable improvement in effectiveness and efficiency. The figures are valid but they are not realized in full-scale implementation. The problem is rarely the technology. It is more usually "behavioral," which really means inappropriate design and implementation, lack of attention to human issues, and lack of felt need. The system may seem like a solution for which there is no real problem if there is no clear business message and policy from the top. The introduction of new office technology has been marked by organizational tinkering. Unfortunately, firms cannot tinker their way into major organizational change and consequent competitive advantage.

The opportunity is there, given vision, policy, and architecture. There are three broad areas where telecommunications improves internal communication: electronic mail, videoconferencing, and document interchange.

Electronic Mail.   The memo or substitute for a phone call is sent from the terminal into a remote computer. It is held there and forwarded to the recipient when that person logs onto the system. He or she may be in Bangkok, London, or en route to Germany. It may be 7:00 A.M. in San Francisco when the message is sent, which means the office is closed in all of those countries. It makes no difference. The message is stored and forwarded.

Electronic mail can transform operations in international firms. High-technology firms on the West Coast are never open when their Asian component suppliers and their European sales office are. Owens-Corning's calculation of the value of computer-based messaging seems a reliable one. It estimates that it saves about 25 cents a message on a market basket of one memo, three local phone calls, and one long-distance call plus around $4 in employee time per message (about 60 percent of phone calls are not successful, because of busy signals or because the intended recipient is out of the office).

Videoconferencing. In electronic meetings, people can get together, often at very short notice, without having to travel. Such meetings are televised discussions, or if "slow scan" and "freeze-frame" methods are used to reduce the telecommunications bandwidth, they are meetings with photographs that are refreshed periodically.

There are other, nonvideo forms of computer conferencing. The simplest extends electronic mail to cover group communication on line. Procter & Gamble has operated its CONFER system since 1978. This allows people who do not work together but need to keep in touch or cooperate on ad hoc projects to carry on "meetings" often spread over weeks or even months.

Audioconferencing is an extension of conference calls. Procter & Gamble, among several others, has an executive room that allows board members who cannot get to a meeting but whose comments are needed to participate almost as if they were there.

Document Interchange. This can be summarized as paperless bureaucracy. It is not the paperless office that early enthusiasts of office automation promised; that has not come to be and in fact better tools for word processing have increased paper since people now make far more corrections, get very fussy about format and appearance of documents, and like to work from clean copy rather than cut and paste and add handwritten annotations and interim changes.

An early illustration of document interchange across organizations is Motornet, a system launched in June 1985, after an extended pilot, by a group of car manufacturers and their suppliers and subassemblers. The network is based on United Nations guidelines on document formats for international trade: purchase orders, invoices, shipping bills, and so on. These standards, called Simplification of International Trade Procedures (SITPRO), have existed for some years. It took the business opportunity of cutting out paperwork and the time, errors in handling and delays, and costs they create plus the mergence of operating systems for document interchange to get them used.

These are the main areas of opportunity. It is clear that they offer many practical and proven benefits. The examples below show that.

Some Examples of Benefits

Texas Instruments (TI) has evolved its international electronic mail system over 20 years, always planning ahead to handle multiple applications. Its network interconnects 13,000 terminals and a number of small and large computers. It carries voice and data over the same links. The system handles close to 1 million transactions a day, with a maximum response time of 6 seconds, worldwide.

TI's cost per message for electronic mail is less than 4 cents, again worldwide, and that for processing an on-line query into its on-line reporting system is under 3 cents. This is a measure of the value not just of electronic messages versus mail and phone, but of the multiservice network that has a clear architecture. It has been able to expand capacity, add nodes, and upgrade the technology smoothly. One of the reasons the unit cost is so low is that the company has been able to exploit economies of scale and efficiency.

For videoconferencing the evidence in most firms is more mixed, with many pilots, some successful limited applications, but only a few striking organizationwide successes. The main business justification has generally been the anticipated savings in airfares and hotel costs from substituting telecommunications for travel. These have often not materialized. People add meetings, rather than cut travel. The real benefits relate far more to more effective use of time, the ability to respond to problem-solving situations, reduced training costs through courses where the students do not have to go to the classroom but instead it comes to them, and improved coordination of projects that involve multiple sites.

Videoconferencing is expensive. It requires much higher capacity transmission links than other standard applications, and special rooms. The standard architectures are not designed to handle the extra demands it places on the network. The entry fee for even a limited pilot is around $500,000. Cheaper variants that do not provide full-motion video are available, but they have tended to be very limited in scope and quality and undramatic in impact.

Among the successes in videoconferencing is J.C. Penney, which installed a link between its New York and Dallas offices in 1984 to let buyers show merchandise to each other. Since then, the system has been expanded so that accounting groups, real estate planners working on modernizing Penney's 1,600 stores, and marketing planners can all meet electronically. Displaying goods selected by wholesale buyers to merchandise managers at 175 viewing sites in the United States, the system eliminates the previous process that entailed over a dozen trips a year to New York headquarters for 65 buyers from sundry regional offices.

The new process saved J.C. Penney over $4 million in sample inventory and $500,000 in travel costs in 1986 alone. Furthermore, Ross Longyear, president of J.C. Penney Communications, Inc., claims that the primary benefit of the system is "to improve our communications in the merchandise-buying process to ultimately gain a competitive edge. Stores are buying what they think will sell rather than get what somebody else had chosen for them" (*Communication Week*, March 23, 1987, p. 38; by Saroja Girishankar). The savings in direct travel costs have covered half the total telecommunications bill. The indirect savings are far larger, even if not easy to measure: travel time, fatigue, and another variant on information float: "We'll have to leave that to be decided at our next meeting."

Boeing was able to beat its original schedule for developing its 757 airplane at least partly because it had linked its scattered facilities in Seattle through four videoconferencing locations. Some of these are almost 40 miles apart. Managers, engineers, and pilots were able to make instant design decisions.

Project management can be improved through videoconferencing. NASA, Bank of America, Aetna, and Hughes Aircraft, as well as Boeing, testify to its allowing teams working at different locations to "meet" more often in short meetings without losing track of details and progress. A study by Ford of Europe showed that participants in a one-hour video meeting accomplished what usually took four hours. They also reported many highly effective 15-minute meetings, which would have been impractical any other way.

American Airlines institutionalized audioconferencing as part of its daily operations. The meeting is attended by corporate managers, airport staff, and even personnel traveling in cars and on planes at the time. It averages less than 20 minutes. (From Robert Johansen, *Teleconferencing and Beyond*, McGraw-Hill, 1984, pp. 60–61.)

Document interchange is among the most complex aspects of telecommunications, since it entails exact reproduction of format as well as content (the layout of an invoice, for example, or the line items on a bill of lading). This requires far more advanced operating system software, which has only recently become available, is often still unproven, and is highly vendor-specific.

Today, the success of the manufacturing industry depends to a significant degree on its ability to use technology in product development and manufacturing. A December 1984 report on the state of computer-integrated manufacturing points out that the typical manufacturer has 5–35 different types of part number, 3–10 types of bill of material, and 6–20 types of cost. It will also have 16 types of schedules and 7 types of change to them. In 1988 production lines are even more intricate, and the implementation of CIM is critical to a company's future. One engineer commented that modern manufacturing is so complex that no one person can collect all the whereases and wherefores in his head.

While establishing a telecommunications network is increasingly a make-or-break factor to a company, many of the major payoffs come from inter- rather than intracompany communications; however, standards are still needed, and it will be years before they are fully in place, although the X.12 international standard is rapidly emerging as the base for cooperative agreements within an industry. (The grocery and railroad industries are leading examples.)

It is mainly when one company has the clout to force others to adopt a standard, or at least to gently suggest that cooperation may be in their economic interests, that order comes from chaos. General Motors with MAP, IBM with its Document Interchange/Content Architectures (DIA/DCA), and AHS are examples of this.

Sometimes, cooperation comes from recognition of joint interests. The Society for International Worldwide Funds Transfers (SWIFT) is an illustration. Competing banks need to communicate. The SWIFT format for funds transfer is a cooperative arrangement that in no way reduces competition but strengthens the whole industry. In 1977, 400 banks were on the network; in 1987 this had grown to over 2,200. SWIFT planned to have its long-awaited new system SWIFT II in place to accommodate the sky-rocketing demand for the network. Countries scheduled to be connected in 1988 include Bahrain, Columbia, Tunisia, Panama, and Kuwait.

Despite SWIFT's size and popularity the system is not without its competitors. Several big American banks have developed their own systems and potential SWIFT II members, upset by delays, are beginning to wonder if SWIFT should be their network at all. Ironically, it has also opened up threats to it. Ford saw that if it, too, adopted the SWIFT format for its internal money transfers via its own network and became a member of SWIFT, it could bypass its banks.

Owning a standard can create barriers to entry. This has been true especially for IBM. Sharing standards eliminates barriers to entry. This has also been true with IBM. Firms like Hitachi, Compaq, and Amdahl created their market position through adopting IBM standards. Every vendor has been pushed toward accepting IBM's telecommunications standards.

Electronic document interchange between firms will make the issue of standards even more volatile. The strategic business question it raises is When do we cooperate and when do we compete?

EXECUTIVE INFORMATION SYSTEMS

Computerized MIS in most firms provide the worst common denominator of detail in the reports they create. They are too detailed for managers to spot trends easily and too disaggregated to help them pinpoint exceptions and track down specific items of interest or concern.

In addition, the information senior managers get is generally based on historical financial reports; the information float they reflect is far too long for executives to be able to anticipate problems rather than react to them when they have become apparent, which may be some time after they actually happened. For example, an increase in inventories reported in September may be the outcome of increased price competition in a particular market in July.

Management information systems have been built around the financial reporting cycle because this provides the most standardized, and best established pathways for moving information across the firm. It is the mule train over the mountains: reliable, but slow and cumbersome. Unfortunately, many firms use the new electronic highways to carry very much the same information. On-line access to the same old reports is just mules on amphetamines.

In companies that have rethought their information needs, a new style of "executive information system" (EIS) has emerged, exploiting centralization-with-decentralization. These systems

- Pull key operating indicators to the center
- Are based on disaggregated data stored in the business units
- Are presented in meaningfully aggregated form, often graphic displays of trends
- Allow the user to work back to the disaggregated data to track down particular items.

An EIS fails, however, when there is no clear idea of what to do with the information. Putting useless data on-line does not make it useful. Managers will not want to waste time accessing data that does not add value to their time and quality of decision-making.

The successful EIS is driven by its senior manager sponsor. An example is Banco Internacional de Columbia's president, Michael Jensen, whose EIS generates a daily balance sheet and income statement. At first sight, this may seem a little silly: What meaning could such reports possibly have?

Jensen used the daily reports as a template and frame of reference. He was looking for trends and potential problems and the daily statements provide a framework for this. His comments capture the nature and value of the system:

> The questions for the system are limited only by my experience and imagination. I have preformatted questions that I ask every day. For example: "Give me any overdraft for more than 3 million pesos, overdrawn for more than 10 days."
>
> All this information (multiyear "tendencies" and "long trends") is so that your stomach feels good and we—the Board of Directors and myself—can sleep at night. We need a snapshot of trends.
>
> What is critical for me to know? For my managers? (Comments from personal communication with author.)

Jensen gained a far better sense of control than before. He could feel on top of things and never more than a day behind in knowing where the firm stood and was moving.

What happens when the CEO goes on-line has been reported in detail by John Rockart and Michael Treacy:

> Ben Heineman, president and chief executive of Northwest Industries, spends a few hours every day at a computer terminal in his office. Heineman accesses reports on each of his operating companies and carries out original analyses using a vast store of data and an easy-to-use computer language. The terminal has become his most important tool for monitoring and planning activities.
>
> George N. Natsopolous, president of Thermo Electron, writes programs in the APL language to format data contained in several of his company's databases. As a result, he can quickly study information about company, market, and economic conditions whenever he desires. (John Rockart and Michael Treacy, "The CEO Goes Online," *Harvard Business Review*, January/February 1982: 82.)

Obviously, this opportunity to exploit the firm's existing data and communications system is one that senior executives in fast-moving, dispersed, and complex businesses can hardly afford to neglect. They must actively choose it, however, An effective executive information system will not be created as a standard part of the data processing department's activities.

Since Rockart and Treacy wrote about the experiences of senior managers in making use of EIS, there have been plenty of unsuccessful attempts to create them. The main reason for failure is passive endorsement by the senior executives ("Give me an on-line system") with no clear idea of what information they really will use if provided and what they will do with it.

COMMITMENT AT THE TOP

One factor clearly distinguishes the companies that have been able to harness telecommunications, data, and computers to make a difference to how the firm operates. It is that the systems address some core aspect of the operations, and the development is driven by a business priority. People care—the management sponsor, the users, and the technical staff all have a clear idea of why this is needed and what it means. The technical vehicle is the means to a recognized and focused end. The designers will often make technical mistakes, and there are invariably managerial and organizational problems in making the innovation take root. But these are just roadblocks to be moved. They get removed because there is a momentum for change and change offers real benefits.

It is one of the cliches of the computer field that effective introduction of information technology requires top management "commitment." That commitment is often, though, just a memo of approval. The best commitment is behavior, not words. In the most effective examples of using telecommunications, top managers have taken charge of change. How they can do this, systematically and simply, is the topic of Part IV of this book. This chapter ends with one of the best examples of how an organizational vision can and should drive the use of telecommunications.

Hercules' Integrated Network

Hercules is a chemicals firm, headquartered in Wilmington, Delaware. Its sales are over $3 billion. It sells over 1,000 products and operates 40 domestic plants and 24 sales offices. Of its 26,000 employees, in 80 locations worldwide, only 10 percent work in the Wilmington area.

In 1979 the president of Hercules, Alexander Giacco, initiated a study on the firm's organization. The priorities were to find ways of linking Hercules' locations more closely, improving professional and managerial productivity, and exploiting the opportunities opened up by the move into a new corporate headquarters to rethink how the firm was to be run.

As a result of the study, Hercules developed its Automated Office Systems (AOS) Program and signed a long-term contract with Satellite Business Systems to install and operate an integrated network that would carry voice, data, and images along the same high-speed transmission path. The network was a fixed-cost operation, a private network with plenty of spare capacity. The basic philosophy "was to overbuild, rather than underbuild, and then to cut back later if necessary."

The main applications reflected the president's strongly stated wish to shorten lines of communication and keep the people who do the work, who are involved in the planning, and who have key responsibilities, in direct contact with one another. In 1987 Hercules held between 50 and 60 video meetings each month in 25 videoconferencing rooms (5 at headquarters, 20 in other domestic and international offices). Videoconferencing, the managers report, "not only saves time and travel, but forces us to make the meetings more organized." The frequent meetings give them "a chance to exchange information that otherwise would wait."

Hercules' videoconferencing rooms paid for themselves in two years by saving the company travel-related expenses and increasing productivity. Another application is the Voice Message Service. Voice messages, like electronic mail, are stored and forwarded, saving time in reaching sales representatives or busy plant managers. The Voice Message System allows people to get messages quickly and gives them flexibility, since they "can call any time of the day or night." Serving approximately 3,500 users, the system issues a monthly average of 74,000 messages. A third application, routing text electronically around Hercules' offices worldwide, has reduced secretarial costs by 40 percent, a savings of $3 million a year.

Hercules has cut out six to seven entire layers of management. The key to making all this work has been the commitment and vision of the senior managers as well as the firm's president. They exemplify the observation of Ross Watson, head of the Information Resource Department, that top managers "must sign on and be the real leaders of cultural change if there are to be results."

5 MARKET INNOVATION

STEALING A MARCH

The most profitable application of telecommunications occurs when a firm uses it to steal a march on its competition by moving into new industries, intruding on other ones, or repositioning its business.

Preemptive Strikes.   Certain business moves create such a level of customer acceptance that they both change the competitive dynamics of the marketplace and push other companies into making a response. If an innovation is easily copied, it is not preemptive. Telecommunications enables the preemptive strike because the costs, lead time, and long learning curve needed for managers and technical personnel to develop a large-scale network create a barrier to imitation. Regardless of the degree of competitive advantage to the leader, the followers are at an immediate competitive disadvantage that lasts for years.

Repositioning the Business.   Building a new delivery base can open up major market opportunities. Getting there early creates the base for a stream of new products, which together push competitors into a defensive position. Dun and Bradstreet, Reuters, and Citibank are successful examples.

Piggybacking.   Firms that already have an electronic network can add new types of traffic to it at relatively low cost. If they can interconnect their networks to other firms, they can create not just a new product, but in some cases a new industry. Electronic funds transfer at point of sale and electronic data interchange are obvious examples.

Deregulation can fuel the process by allowing firms to step across traditional boundaries between industries.

Discussions about the ability of firms to use information technology to make radical business innovations leading to a sustainable competitive edge swing from boom to gloom. For several years, beginning around 1983, a groundswell of enthusiasm grew for the argument, with four examples constantly being cited:

- American Hospital Supply, which changed forever the value-added chain in distribution via electronic delivery links and by locking in the customer through communications-intensive services

- Citibank, which became the reference standard for innovation through the much publicized aggressive drive of John Reid, under the truly visionary leadership of Walter Wriston, to make Citibank the leader in electronic banking and credit cards, plus the quieter but in many ways even more radical, and for a decade much more profitable, set of initiatives in international corporate electronic banking led by Tom Theobald

- Merrill Lynch, whose cash management account entirely preempted the banks, from which it effectively removed $80 billion of customer deposits

- American Airlines, which moved from one of the weakest major airlines in 1978 to a preeminence it has sustained through the 1980s, with SABRE, its reservation system, the cornerstone to its success.

Every fashion generates a natural backlash—of boredom, contrary commentators, or simply the need for journalists to find a new perspective. Much of the old enthusiasm looks bombastic now, or even naive. Merrill Lynch's economic and managerial performance has been less than stellar since the mid-1980s. AHS is now absorbed into Baxter Travenol. Citibank's much-vaunted culture of confidence, expansion, and arrogance has turned introspective and less assertive. American Airlines (AA) still holds its position, but there are many signs that AA will not have an easy time maintaining the profit flow SABRE has provided; it makes more profit than AA gets from flying airplanes.

Some commentators accept that all these firms got a competitive edge but at great expense and only for a short time. They also see them as one-time events: lucky in their timing and nonrepeatable as the vacuum is now filled.

This chapter faces up to the challenge of showing that telecommunications is still a major force for market innovation—there are plenty of vacuums left—and that the edge can be sustained. Two important points to be made right at the start, though, are:

- No edge is sustainable forever—not patents, products, management, or market control. Very roughly, a rule of seven seems to hold; it takes about seven years to make major business innovations involving major investments in the technical infrastructure, and this gives a seven-year window of opportunity before the advantage erodes.

- "Can be" does not mean "will be" sustained. Merrill Lynch, in particular, failed to build a product stream on its original CMA base and failed to match its sales force incentive system to its strategy. American Hospital Supply kept its edge until the social pressures and economics of the medical industry made life difficult for the best and the weakest alike.

That said, the three strategies for using telecommunications for profitable market innovation identified in the 1986 first edition of this book remain just as valid today, and when every large firm has access to the same technology, it is the management vision, courage, speed of response, and ability to move the culture with the strategy, that make the difference between winning and losing.

Who Is Our Competitor Now?

Who would have guessed in 1976 that within five years a bank (Citibank) would be in head-on competition with a securities firm (Merrill Lynch), a vendor of travelers' checks (American Express), and a department store (Sears) to create the "financial supermarket"? The object of each was to integrate a range of financial products under one roof, within one customer database, and at one electronic delivery point.

Similarly, who would have forecast that a long-established news service (Reuters) would become transformed into the world's leading provider of international financial data, or that it would be retailers and oil companies, not banks, who controlled EFT/POS in the United States, or that companies like GEISCO would move from being suppliers of telecommunications services to firms that used them to sell such products as electronic cash management to being their direct competitors, marketing the same products?

There is hardly a single industry that can ignore the threat from or the need to cooperate with firms outside its own. Regardless of the level and sustainability of the competitive edge, business innovation via telecommunications makes the word "industry" meaningless and also adds to the question "Where do we lead, where do we follow?" a related one, "Where do we cooperate, where do we compete?" With telecommunications, nonindustry players become direct competitors, customers become competitors, and new strategic alliances—more than contracts but less than joint ventures—become desirable, and rather dangerous, too.

Non-industry Competition.   The airlines are struggling to retain the control of their own selling outlets and process. American Express is a major threat via its card base, travel management services, and massive telecommunications infrastructure. Dun and Bradstreet is at least as big a threat, by its linking of its on-line Official Airlines Guide to its travel agent chain.

Merrill Lynch did a lot of damage to the banking industry through its Cash Management Account, which gave it a solid base of one million upscale customers across the broadest range of services in the industry. Had its strategic alliance with IBM worked out, it would also have intruded on financial trading and information suppliers. Luckily for the competition, Merrill Lynch and IBM both found it hard to work together and the IMNET system they created was clumsy, slow, and not worth the $250 million wasted investment.

Customers Become Competitors.   In the early 1980s Ford applied to become a member of SWIFT, so that it could move funds around through its own network. Texaco similarly bought 50 percent of a London foreign exchange dealing firm, so that it could bypass banks in managing its $3 billion of cash. British Petroleum set up its own in-house bank, to handle its own finances first but then to sell services to outsiders. Volvo is also now an in-house bank, as dozens of multinationals will soon be. At some stage they will start using their advantage of cash flow and funds without the bank's disadvantage of having to buy deposits and will move into some form of selling services and financing.

New Alliances.   Any firm that wants to capture part of the customer relationship via electronic delivery needs a credit card. Does it do this by issuing its own, branding a Visa card, or simply accepting standard cards? Sears' Discover card is an example of the decision to compete, not cooperate. Merrill Lynch's CMA is in fact a Visa card, with Banc One of Ohio providing the processing base—thus Merrill Lynch has chosen to cooperate, not compete.

For retailers and oil firms moving into point of sale, the simplest option is to accept Visa, Mastercharge, and Amex. Several of them, though, recognize the long-term strategic issues that affect the decision whether to cooperate or compete:

1.   Whose customer is it?
2.   Who owns the customer data?
3.   Who controls the pace and nature of future market developments?
4.   Is one of the allies able to use the other's product as its own loss leader to create an incentive for the customer to use the card, to the advantage of both?

The last point is a relatively new strategic element in credit card providers' battle to gain "wallet" (rather than market) share. Citibank and Dreyfus provide early warning signals here. In mid-1987 Citibank issued a Mastercharge card that offers one frequent-flyer mile on American Airlines for every dollar of expenditures charged to the card. Citibank pays American two cents per mile. The cooperative arrangement uses the airline's product as the bank's loss leader. Both of these organizations fully understand the critical value of owning the resulting customer data; they have apparently agreed to share it. Citibank benefits immediately and immensely; the usage per month of the card is several times higher than a standard one. American obviously increases its hold on frequent flyers.

This alliance works to the advantage of both parties. Dreyfus's use of another industry's product similarly creates an incentive to use the card but to the disadvantage of the other side of the transaction. Dreyfus will cover the collision damage waiver (CDW) insurance for car rentals when the card is used.

The CDW is extremely profitable for car rental firms and the price charged is totally unjustifiable in actuarial terms. Customers realize they are being gouged but in many instances feel they have no choice but to take the coverage. Note that Dreyfus is not offering coverage at a lower price; it does not seek to compete with the rental firm's product, only to use it as its own loss leader.

The need for and risks in cross-industry alliances or raids will increase and increase now that credit cards are central elements in the strategies of banks, department stores, airlines, gas companies, traditional card providers, and even car manufacturers (Subaru offers its own branded Visa card to certain car buyers), and as customer data becomes a crucial tool for market segmentation, cross-selling, and increasing customer loyalty.

These are questions that quite literally are determining the shape of whole industries, subtly but surely. For airlines, retailers, petrochemicals, banks and insurers, they are already urgent. The difficulty is that firms have to make decisions now to be in position years from now. The questions relate to control of a firm's, or even an industry's, outlets and distribution system, retention of customer loyalty, options for innovation, and long-term cost base.

Preemptive Strikes: Imagination Plus Technology

American Airlines and United shook up the airlines industry by being first to seize the opportunity of telecommunications. American spent $200 million on SABRE, its telecommunications reservation system, and in 1988 has over 40 percent of the total travel agent computerized reservations business. United spent $150 million and has over 30 percent of the market. Delta, TWA, and Eastern share most of the 25 percent of the travel agent market left by American and United.

That was only the first threat. In 1981 American preempted the entire industry with its AAdvantage program, the first frequent flier bonus program. At one level, the AAdvantage program looks like a marketing ploy, an effort to build brand loyalty and get a short-term increase in market share. Since the other airlines would have to match it, its competitive impact would be limited.

In 1982 similar bonus schemes in the car rental business failed badly. Avis made the first move. Hertz and Budget had to respond. Only the customer gained. Two years later, after the programs had been ended, the three firms' market shares were almost exactly the same as they had been at the start. The free gifts did not lead people to rent cars more often. They did add cost and sometimes near chaos to operations. The check-in counters looked like department store displays. There must be many business travelers who still own a dozen or so Pierre Cardin calculators (Hertz), a few of Avis's distinctive beige and red travel bags, and perhaps a teddy bear.

The American Airlines initiative was not a simple promotional scheme that could be copied in a month or so, though. It was the result of a seven-year development that rested on technology, not marketing. American's strategic goal was to capture in its reservation system database information on the 400,000 people who made up 70 percent of full-fare travel, and then to use that as a cornerstone for service, marketing, and product development. American's competitors did not have the database, and many of them did not have the reservation infrastructure either. They got all the disadvantages of the program, though. American's reservation system automatically keeps track of members' flights and bonus mileage, because the database is organized to provide a history and profile of each of them:

> The attendants knew that William George was in seat 4A. The reservations computer, communicating through a terminal in the cockpit, had alerted them that Mr. George is an American AAdvantage Gold member—in other words, a super-frequent flier to be coddled by name. Had it been afternoon, the computer would have indicated his favorite drink. (Andrew H. Malcolm, "On a Wing and a Computer," *New York Times Magazine*, February 12, 1984, pp. 16-23.)

Since they did not start off with the database linked into the telecommunications reservation system, the other airlines had to set up complete separate systems, which were costly to run. One of them, for instance, provided frequent fliers with a booklet of preprinted slips. These have the user's identification number in computer-readable form. The traveler fills in the flight number. Unfortunately, the ink and handwriting are not machine-readable, so that this airline had to add an estimated 40 full-time staff to key in the data. Another airline provided stick-on labels, which were machine-peelable, so here again clumsy and expensive procedures had to be added to handle "data capture." Neither airline has been able to link the data into its reservation system.

American continues to build on its advantage. By the end of 1985 it had installed automatic ticketing machines in 45 airports and city ticket offices. These TransAAction centers allow travelers to make reservations up to 300 days in advance. American is a leader in telemarketing, using the spare phone reservation capacity to run telephone surveys and sell programs for itself and outside firms. It was clearly getting ready at the end of 1985 to move as aggressively into the European travel agency market as it had in the United States. The 1987 airline alliance, led by British Airways and United Airlines, set up to build a new international reservation system, was to a large degree motivated by their concern about this.

Discussions of how firms in the airline industry have responded to the increasingly competitive environment following deregulation single out the way American has made computers and telecommunications almost an everyday part of its business thinking. As Alfred Norling of Kidder Peabody observed in 1984 concerning this response, "no one has been as imaginative, as tough, and as successful, so far, as American Airlines."

Response from the Competition

The competition did not, of course, stand still. Since 1984, Computerized Reservation Systems (CRS) have been belatedly recognized as an issue as important in competitive positioning as route structures, and CRS and "marketing" and "distribution" have become almost interchangeable terms. Leading players have had to make numerous moves and countermoves.

When Carl Icahn acquired TWA, his known original intent was to hive off PARS, TWA's CRS, which was technically the best in the U.S. industry, into a separate company and get rid of the airline. He offered PARS to IATA, the international airline association, when IATA was fairly frenetically trying to create Neutral International Booking System (NIBS) to counter SABRE. TWA's asking price was rumored to be around $750 million and IATA's offer only $450 million. No deal was made.

Frank Lorenzo's acquisition of Eastern, in disastrous shape as an airline, was because his assembly of airline companies lacked a CRS. As did Icahn, he spent considerable time and money trying to work out what premium Eastern's reservation system was worth. The answer seems to have been around $400 million. The Northwest-Republic merger similarly centered around CRS as well as routes and shared economies of operations.

People Express's demise partly came because it lacked a sophisticated CRS to handle the complexity of fare structures and hubbing in the turbulent deregulated gladiatorial arena of the mid-1980s. This contributed to establishing Continental, part of the Lorenzo megacarrier which includes People Express, as an airline whose concept of service, scheduling, and efficiency kept it at the head of the consumer complaint list as the U.S. answer to Aeroflot. The People Express staff were untrained to handle a computerized check-in procedure case. Ironically, People Express had been imaginative and effective earlier in its installation of a pushbutton direct reservation system that bypassed reservation agents. That worked in a simpler era of business, with fewer routes and stable pricing. (American Airlines makes 106,000 price changes a *day*.)

British Airways' resurgence from an airline within 48 hours of bankruptcy and from a reputation for service whose quality and style used to match that of British cooking to becoming one of the premium international airlines is as remarkable as Chrysler's turnaround. British Airways had been a pioneer in CRS in the 1970s and in 1987 it set up a joint venture with United Airlines, KLM, Swiss Air, plus several other European airlines to create Galileo, a new generation of CRS designed for the new marketing, selling, and operating environment of the 1990s. With the possible exception of Scandinavian Airlines, no other European carrier has so clearly grasped the critical importance of CRS to international competition.

Another consortium of European airlines refused to join Galileo and set up its own joint venture to create a rival system, Amadeus. It then faced the problem that BA and its partners well understood, that it had to find a U.S. partner. Here that could mean caving in to SABRE, the Trojan horse that they all now fear. In fact, the consortium came close to a deal with SABRE. Instead, Amadeus chose—or had to choose—Eastern's SystemOne. Amadeus seems at least partly driven by Lufthansa's refusal to be part of a consortium where BA plays a leading role and by the purely technical (and from a business viewpoint not demonstrably justified) wish of some of its partners to move from a Sperry-based CRS to an IBM one.

Eastern and Northwest in 1987 moved aggressively to try and displace SABRE, through incentives and price cuts, even agreeing to cover legal costs in any disputes over broken contracts.

American Airlines caused all this. In every single instance listed above, the overt or covert issue was responding to, cooperating with, or displacing SABRE. All in all, it is impossible to understand the competitive moves of domestic and international airlines without recognizing just what SABRE gave American and what it still does. The system was key to gaining occupancy in the travel agent business, to launching and differentiating the AAdvantage frequent flier program, and to creating AA's hubbing strategy. "Hubbing" (rationalizing routes and schedules) has been a central aspect of postderegulation strategies. AA did not invent hubbing but became the acknowledged expert. SABRE provided it with the on-line passenger and schedule data to do so.

The comments of Robert Crandall in 1986 to Boston financial analysts are not an exaggeration:

> As changes evolve, we intend to work hard to maintain our leadership position; moreover, the strengths we've developed in the travel business open some potential opportunities in other industries. SABRE has the ability to handle very large volumes of complex transactions which depend upon the storage and manipulation of huge proprietary data bases and the kind of quick response that SABRE produces is important to any business with perishable inventories and short-lived transactions. We think that as more and more companies seek to tap national retail markets, our unique on-line capabilities, together with our nationwide communication network, should provide new diversification opportunities in marketing and distribution.

United Airline's rather more sorry story can be used as a cautionary parallel to American's. At the start of the 1980s UAL's Apollo system was the equal of SABRE in terms of penetration of the travel agent business. UAL had the stronger balance sheet, stronger route structure, and larger market share. Something went wrong on the way to the future. Richard Ferris, UAL's CEO, had a very clear vision of the electronic marketplace.

> More important (than marketing "synergy"), says UAL chairman Richard Ferris, are the potential operation links. Hertz, like United and Westin, sells a perishable consumer good. An unsold seat or bed or an unrented automobile is lost forever; none of them can be put in inventory.
>
> It is the airlines—especially United, with its Apollo reservations system, and key competitor American Airlines—that lead the way in using computers to reduce empty seats or rooms or idle automobiles through timely fare or rate discounts. The synergy, in other words, comes through shared communications. ("Ready for Take-off," *Forbes*, August 12, 1985, pp. 30–31.)

United Airlines' business strategy was very different from American's. It was emphasizing growth through acquisition, while American was relying on internal growth. United had the same key business asset as American, a reservation system in almost as many travel agents' offices. It was probably even stronger than American in some aspects of use of office technology and cost-effectiveness of its tele-communications and computer operations. But it had not shown any-where near the same business imagination and speed of response to the changing market as American. *Forbes* magazine summarized its position in late 1985 as "calamity's child," dogged by labor problems, a pilot's strike, and expensive delays in expanding its fleet.

In 1987 Ferris lost control of UAL, which had just been renamed Allegis. He lost the confidence of his pilots (who ended up buying the airline), Wall Street, his own board, and his managers. United has gone back to being an airline, selling off all the major nonairline assets, including Westin hotel and Hertz.

Ferris had a business vision. Visions are not contagious. However sound his stategy was, he failed to move the culture with him. Time alone will tell whether or not the pilots were right or whether they snatched defeat from the jaws of potential victory. In any case, SABRE set the agenda for the airlines in the 1980s. The irony is that it is not the best CRS in terms of technical design. TWA's PARS is better. So what?

The Alien Invader

At least American and United are in the same industry. Perhaps the classic preemptive strike based on telecommunications is Merrill Lynch's CMA, which took the major banks entirely by surprise. It also took away a lot of their deposits. CMA integrates a wide range of financial services: checking and deposit accounts, a Visa credit card, and a securities account. Money is moved among them as needed. There is no float, and charges are made only when a transaction (such as selling stock) is completed, not when it is initiated. Another CMA innovation was to pay money market interest rates on checking account balances.

In its first 12 months of operation, CMA brought in $5 billion of funds to Merrill Lynch. The minimum required deposit was $20,000. By 1984 the firm had one million CMA accounts with balances of $70 billion. Merrill's strategy was a simple one: Open as many accounts as possible and make sure there is plenty of computer and communications capacity to handle the volumes.

The CEO of one of the top five banks promised that his firm would have a CMA in six months. It did not have one three years later. Merrill had picked up a long lead on the banks in their own territory.

None of the counters to CMA, such as American Express and Shearson's package, has had anywhere near the same impact.

As with the AAdvantage program, the banks had to respond. They did not have the technology base to do so. They could offer special interest rates or "personal" banking services but could not provide the processing and database that is the key to CMA.

The technology Merrill Lynch used was fairly standard. There was no state-of-the-art software or hardware. Merrill used Banc One of Columbus, Ohio, to handle its processing. What was special was the way the systems were packaged together to create the distinctive business product that neither Citibank nor American Express could match, even though they were spending $300–500 million on information technology: no float, instant movement of funds, and a single access point to multiple service.

In retrospect, it is still not clear what the banks could have done when CMA hit them. Some of their efforts were little more than advertising gimmicks. Merrill Lynch established itself by this one move as a central player in banking. It has over a million CMA customers. That means it has positioned itself as a focal point for all the financial services needed by affluent Americans.

It is still not entirely clear how much Merrill benefited from CMA, largely because it got into trouble in its traditional business of securities broking, especially in managing and motivating its sales force. In addition, Merrill did not create Son of CMA, a product stream based on its original CMA that would have made the banks' problem of catch-up even harder. That said, CMA caught the banks by surprise, changed their industry for the next decade, and above all raised the question so central to competing in the electronic marketplace: Who is our competitor?

Rethinking Business Assumptions

The outcome is not yet so clear in all examples as in the preceding ones. Citibank's massive commitments to electronic delivery internationally and domestically helped it push Bank of America from co-leader in the industry to a badly faltering also-ran and Chase to a firm in search of a strategy. Citibank's lead looks solid, but the race is still on.

Citibank is especially interesting in the context of telecommunications and business strategy in that so much of its technical investment was poorly made and badly managed. The Citibank story includes many instances of badly designed systems, blunders in project management, waste of resources, and naive gambles on advanced technology which looked silly at the time and were. But the vision was there, and it provided a drive and clarity of purpose that overrode the mistakes. Like American Airlines, Citibank has now set the competitive terms and pace for the rest.

Of course, having made a preemptive strike is not enough to maintain the lead. It is hard to keep ahead. The companies that were the exemplars of 1982 can be the flops of 1988. Several of the firms the authors of *In Pursuit of Excellence*, Peters and Waterman, picked out as excellent soon became examples of the opposite. Atari is one instance.

Even when they have faltered, though, the companies that built the highways maintain a strategic edge. AHS still has its vast electronic distribution network that the president of Baxter Travenol thought could add $400 million in sales a year if his firm were to merge with AHS. Merrill still has the $70 billion of funds brought in by CMA, plus a direct and broad financial relationship with a million affluent and thrifty Americans.

Scanning the Competition for a Preemptive Strike

It is always hard to make things work in the information systems field, and all the components needed for the preemptive strikes described took a long time to design and implement, but they did not break new ground. There was nothing here that was not and is not well within the capabilities of any well-run large firm.

The innovation is in the business thinking. Each of the firms had a clear idea of where the market was moving. They had a target of opportunity. Many of their competitors recognized the importance of telecommunications and invested heavily and sometimes more effectively. Chase's funds transfer network, for example, was far more efficient than Citibank's. United was American's equal in reservations. They did not, though, have a driving business vision. Their approach was diffuse in terms of business concept, market strategy, and perception of the customer. Technology is an enabler. It cannot create a business strategy.

Apart from a clear vision, the successful preemptive moves had to have an early, high-level commitment from management. They could not be cost-justified. They had to be centrally funded and required a business act of faith, both because of the five- to seven-year lead time each involved and the problem of forecasting volumes and revenues. CMA exceeded its five-year target well within the first year.

One wonders in how many other companies around this time a business planner or imaginative data processing or telecommunications manager was thinking along the same lines as Merrill Lynch or American or AHS, but nothing came of the idea because "How do you cost-justify something like this?" You don't.

No large firm can afford not to try to anticipate a preemptive strike from within or outside the industry and to defend against it. The strike may come from any major firm that is aggressive in its general business strategy; that is looking to expand its activities or diversify; and that has a strong telecommunications network, in terms of scale, scope, and customer contact points. It may come from any competitor that built a private network three to five years ago and that uses workstations as a primary contact point by or link to customers. Outside the industry, watch out for firms that keep using the words "integrated services" in their annual reports and advertising in the business press. Part III of this book describes the telecommunications infrastructure needed to guard against others' innovations.

REPOSITIONING THE BUSINESS

Mobilizing the Organization for Change

Many firms are repositioning themselves for the long term through telecommunications. The impact on competition may be gradual. Instead of making a brilliant raid, such as the AAdvantage program, they are mobilizing for a long campaign. Once they seize the bridgehead, they may be hard to displace. Common to the diverse firms who are repositioning themselves in this way are the following characteristics:

- A very long-term view and a willingness to absorb costs for some years
- An emphasis on keeping control over the telecommunications resource and creating a product stream that exploits it, rather than using other suppliers' networks
- Looking for a vacuum to fill and a market to grow

Dun and Bradstreet exhibits all three of these characteristics. One of its biggest problems in the mid-1980s was how to invest its $650 million of cash and marketable securities. It had no debt and had been able to spend close to $2 billion in acquiring over 30 companies during the previous 18 months. Its return on equity over the past 10 years has averaged over 25 percent.

Obviously, that is an impressive record. It has been built on telecommunications and data. Dun and Bradstreet turned itself into the leader in electronic publishing in just under 10 years. Over half its credit information is now delivered electronically. None of it was in 1978. It acquired Nielsen, the consumer marketing firm, in 1983 and owns Donnelly Marketing, the Official Airlines Guide, and other companies whose business products all revolve around data.

In 1975 Dun and Bradstreet had close to a monopoly in its main lines of business: credit information, printed directories, and airline guides. Then TRW moved into credit reporting, and the airlines installed reservation systems. In late 1979 Dun and Bradstreet decided to become a provider of information services. "Our industry was lying there waiting for the computer age," said Harrington Drake, Dun and Bradstreet's chief executive officer.

It acquired National CSS, a computer time-sharing service. It salted the divisions with computer consultants and market researchers; the business units would drive the product development process, not corporate staff or technical people. The company began a search for new products through "interdisciplinary communication." It added 70 in-house trainers to give courses on working together and learning to listen. Senior management provided seed money to encourage interdepartmental projects so that line managers would not fight about cost allocations and budgets.

It built up its telecommunications capability. DunsNet, its private network, which cost around $20 million, connects customers in over 150 cities directly to Dun and Bradstreet's mainframe computers. DunsVoice, DunsSprint, and DunsPlus are other access vehicles. When Dun and Bradstreet acquired Nielsen, it had the electronic delivery base to augment Nielsen's data resources at little extra cost.

Electronic delivery plus Dun and Bradstreet's continuing investment in software allows the firm to add value to its many databases:

> We've been able to take an existing database, reformat, spin it around, and develop a whole slew of products for our customers.

As long as we viewed this as a mailing list, we were putting blinkers on. We asked what would happen if we said, "This is the most extensive consumer database that exists. One of the things you can do with it is create a mailing list. What other things can you do with it?" The answer became a new division of Dun and Bradstreet: Donnelly Marketing Information Service . . . New products range from consumer spending profiles . . . to services that help retailers and fast-food chains find promising sites for outlets.

"I never thought Nielsen shared Dun and Bradstreet's sense of urgency to be at the forefront of technology," an investment analyst says. Now complacency has vanished. Nielsen is adapting the DunsPlus software to deliver information on consumer goods sales volume, market share, pricing distribution patterns, and inventory levels to personal computers via the DunsNet system. (Direct quotations from "Dun and Bradstreet Redeploys the Riches," *Fortune,* August 19, 1985, pp. 28–33.)

Dun and Bradstreet has been able to charge a premium for data delivered electronically. Its Official Airlines Guide sells for around $150 in printed form and twice that to access via computer network services such as Compuserve.

It has been able to use its technical base to move through the new life cycles of electronic products, from premium innovations to commodity. For instance, in 1984 it wrote off $47 million in phasing out its computer time-sharing service. "The problem with time-sharing is that much of its value-added was shared access to computer power. With PC's [personal computers], the power is in the hands of the customer." ("Dun and Bradstreet Redeploys the Riches," *Fortune,* August 19, 1985, pp. 28–33.)

Dun and Bradstreet has also had to move aggressively to protect its Official Airline Guide's business from being eroded by United and American Airlines selling reservation systems to companies and TWA providing access to its system from personal computers. Dun and Bradstreet bought a leading travel agency chain in early 1985; their rationale: "We'll be able to book reservation systems for travelers so that they'll have an unbiased selection of flights and fares." (And, the spokesman could have added, "We'll get the travel agent commission, not a travel agent or the airline.")

Two points in Dun and Bradstreet's success are very striking. It happened at a time and in a business where many other companies tried and failed badly, and it happened in a firm that had a reputation for being slow and very conservative.

The firms that had failed in the electronic information business have generally been ones who looked for a technological fix to open up a market opportunity or who focused on single product areas. The litany of companies who have poured money into pilots for videotex home information services reads like the ticker tape on the New York Stock Exchange. An anonymous comment in *Business Week* ("Publishers Go Electronic," June 11, 1984, pp. 48–54) characterizes the race—or rather the patrons at the racetrack: "By 1990 electronic publishing will be a $60 billion business. I'm just not sure if it will be $60 billion in revenues or expenses."

It may well have been Dun and Bradstreet's conservatism that has made it such an effective innovator. Its senior managers recognized right at the start that they were "walking a cultural tightrope." They put substantial resources into helping smooth the organizational change that was an intrinsic part of the technical and business change they aimed for.

Dun and Bradstreet avoided one of the most common problems that has blocked firms whose managers have a clear strategy for the electronic marketplace but who ignore the problems of change and thus create a culture clash within the organization. In Citibank, for instance, it took a long time for the "real" bankers, the account officers, to adapt to, understand, believe in, and know how to sell electronic banking. In American Express the rivalries between units have got in the way of its strategy for integrating services. Sears's insurance salesmen in its Allstate subsidiary naturally felt undercut by Sears's moves to bypass them by selling insurance through the mail and electronically. Merrill Lynch's sales force, rewarded on a commission basis, had little incentive to take time to sell many of its new products.

Dun and Bradstreet stands out as one of the very few firms that combined business vision with a sensitivity to the importance of mobilizing the organization behind it. Dun and Bradstreet also had an architecture rather than a set of fragmented application plans. It has been able to add companies to its business portfolio and move in and out of products within the same infrastructure. Nielsen would be a cost-added not value-added acquisition otherwise.

Nielsen provides a small counterexample to the strategy and success of its parent. In late 1987 Nielsen was trying to ward off a preemptive strike, from a relatively small British company that has used telecommunications to change forever the television ratings business.

For decades, Nielsen had a virtual monopoly through its panels of viewers who filled out diaries, on a daily basis, showing which channel they watch. In retrospect, it is astonishing that for so many years Nielsen's primitive diary system has been the basis of setting rates for TV commercials in a $20 billion business and determining which TV shows survived.

AGB is the largest market research firm in Europe and the fourth largest in the world. Its total revenues, mainly from research, are $120 million versus Nielsen's $615 million for 1986. For many years, AGB was one of Britain's fastest growing companies, mainly through the vivid leadership of Sir Bernard Audley, one of its founders. Audley clearly enjoys being Jack looking for a giant to beat up, if not kill. He found one in Nielsen. AGB in 1985 piloted in Boston the use of "people meters" instead of paper diaries to measure electronically and in real-time, rather than after the event, what people were actually watching and to consolidate the information via telecommunications daily.

AGB had been using people meters to capture live data for many years in many countries. Nielsen had been testing them for a decade but had had no impetus to launch them. This reflects both Nielsen's conservatism and its assumption that business would be as usual. *The New York Times* commented (July 26, 1987) that "[Nielsen's] hookup with D&B has not been painless. D&B's strategy revolves around spotting unfilled market needs and developing products to fill them; Nielsen's was based on keeping clients happy." *The Times* describes the impact of AGB's foray: "That orderly world has spun off its axis." Nielsen has had to respond and create its own people meters; both systems went live in the fall of 1987. Initial results showed significantly lower audiences for many TV shows than the diary-based rating report.

AGB came very close to making a preemptive strike. CBS cancelled its long-standing contract with Nielsen. Advertisers made it very clear that they were paying big fees on the basis of ratings information and that that information had better be accurate. The whole market research field has been affected, as consumer goods companies actively look for on-line, timely, and detailed demographic data. Telecommunications-based capture and consolidation mark the accelerating trend. While TV ratings are only a very small part of Nielsen's business, they are core to its reputation. Jack may not have killed the giant, but the giant has a lot of bruises.

That AGB did not quite achieve a preemptive strike mainly reflects its lack of capital and, perhaps more importantly, too long a pilot run in Boston—22 months. Nielsen was able to counterattack; had it been given less notice and time to respond, it would have been in trouble. It has been obvious for years that market research is moving towards online information. The telecommunications technology needed for electronic rating capture is cheap and trivial to install. People meters have been around for over ten years. All the components for innovation were in place in Nielsen, except management awareness and action. It is worth asking why Dun and Bradstreet, so outstanding in moving its culture with its aggressive strategy for the electronic marketplace, was unable to move Nielsen's. Perhaps the answer is simply that entrenched management in a stable market too easily takes the status quo to be the natural future.

Owning the Infrastructure versus Renting It

Dow Jones has run a little behind Dun and Bradstreet but is also a clear success in repositioning itself through telecommunications. It recognized, even more clearly than Dun and Bradstreet, the vital importance of the telecommunications infrastructure. William Dunn, who had been with Dow Jones for nearly 30 years, persuaded his management not to hedge its bets, as had some publishers, by distributing information over other firms' telecommunications networks, even though this was the cheapest and most convenient option, but instead to build a private network. He suspected that "There's more money to be made in distributing other people's information than your own." He said the key to success in electronic publishing is "to be there when the demand takes off."

That means that Dow Jones is in the highway business and lets other firms provide the traffic. It handles 25 other publishers' databases. Because of lead time, Dow Jones, like Dun and Bradstreet, had to build its technical infrastructure and anticipate the business opportunity years before it could expect returns. It had to decide to lead not follow. Dow Jones' News/Retrieval Service began in 1971. It was not profitable until 1983.

Electronic publishing may be to the early 1990s what financial services were to the 1980s. Both are increasingly about turning bits into sales. Both have attracted large investments from firms in very different industries. Both have required cultural change and, in particular, change in the selling process. The winners in these new and risky fields will probably be the firms that have a clear idea of where they want to go and that mesh technology into their culture.

Being There When Demand Takes Off

Reuters is a firm that is already at the convergence of information and financial services. Until the mid-1970s, it was a news service. Like the Associated Press (AP), a company very similar to it but that failed to repositon itself as the market shifted, Reuters relied on slow-speed terminals to send information to subscribers. It had no facilities for them to move information in the other direction.

When the world moved to floating exchange rates in 1976 and interest rates fluctuated rapidly, there was an immense demand for up-to-the-minute financial information. Reuters met it. It expanded its network capabilities and created its Monitor service, which is used by almost every corporate treasurer in international firms. It added a currency and bullion service in 1981 and a dealing service in 1984. In 1985 it set up plans to invest around $65 million in an equities deal-ing service that would allow securities traders to buy and sell shares through Reuters terminals.

At one point Reuters is believed to have come close to making a deal with one of the top U.S. international banks to give it exclusive use of Reuters' network for its electronic cash management systems. At a time when the banks were all racing to get a terminal in the trea-surer's office, Reuters was already there. It had franchised the use of its highway to about 120 bond dealers and could do so if it wished for a bank. It could also add its own traffic.

Reuters was lucky in its timing. It was there when demand took off. As with Dow Jones, an internal telecommunication manager had spotted both an emerging market niche and the key importance of owning the delivery system rather than concentrating on specific applications. As the *Financial Times* commented early in 1985: "On a trading day like yesterday, it hardly matters to the customer how much the product costs or how many components the supplier can provide. What matters is the speed, scope and reliability of the ser-vice." It then added a warning that Reuters will need to prove itself faster off the mark than its U.S. competitors, like Telerate, who are on its tail. Merrill Lynch was slow in creating a product stream from CMA. Dun and Bradstreet has been fast.

The general message here is that while being there when demand takes off and owning the infrastructure is a key to market innovation through telecommunications, so too is exploiting the advantage of lead time. The firm cannot sit back until the competition can catch up and in doing so turn a premium product at a premium price into a commodity. This has happened with electronic cash management, and it could happen with Reuters.

It looks as though it will not, for now at least. In late 1987 Reuters responded quickly to Fidelity's Financial Telecommunications (FITEL) Equinet network, which is a global electronic message system that reduces the time and manpower needed to complete international bond and equity transactions. Equinet links brokers, investors, and financial institutions to a central computer in London over GEISCO's network. The service allows users to compare their copies of an investor's transaction order to the original for accuracy and completeness. FITEL users claim that the system dramatically facilitates and speeds up the previously cumbersome process of moving funds from buyers to seller. (*Network World,* August 10, 1987.)

Reuters entered the competition when Reuters' Holdings of London acquired, in late 1987, Toronto-based Securities Clearing International Corporation and its Instant Link network. Like FITEL, Instant Link is a real-time system that keeps one central transaction order record and allows users to access and update it. As the international financial market grows, so too will competition in this new telecommunications application. And once again the pressure will be on Reuters to move into yet another competitive information technologies venture.

The Network as Value-Added

GEISCO is another firm that intends to be in at the start in as many emerging electronic markets as it can. Its network is its strategy. It is the largest "enhanced communications" facility in the world. GEISCO defines enhanced communication as helping customers get more value out of the computer and data resources they use.

Management Technology magazine summarized GEISCO's strategy for addressing "focused markets."

> Let's say you line up the world and there are 150 vertical markets you could go after. . . . You identify, say, 25 of them where you are uniquely qualified to provide support. Health care, point of sales, and business logistics—from ocean ship-

ping to intercompany trade—are good examples. A third party can help consolidate those businesses. But the only way you can develop a vertical market is to anticipate—show up before they need you.

Industry watchers predicting the convergence of computers and communications are really predicting the past. (Florence Donne, "GEISCO on the Go: The Game Plan of Communications Goliath," *Management Technology*, February 1985, pp. 41–43.)

GEISCO's ability to anticipate has already paid off and it is well positioned for the coming era of electronic trade management: the integration of electronic cash management and electronic movement of trade documents. It has moved quickly into such niches as the "smart building" market, for instance. Its Tenant Services Operation lets tenants of an office building choose from a wide range of communications and information services and access GEISCO's worldwide network. "It was harder for them to sell just their network service alone, but they are the only company bundling these services." (*Banking Technology*, September 1984.)

In 1984 GEISCO won a three-year contract to process payments for the Calwest Automated Clearing House, traditionally the monopoly of the Federal Reserve Bank. GEISCO will be a disturbing force for many firms in many industries. Deregulation allows it to intrude into other firms' territory, but the need for telecommunications makes many of them GEISCO's customers. Owning the infrastructure opens up the business opportunity.

Have Network—Will Travel

Any firm that has a strong communications facility in place and that has the capital to expand has far more opportunity and incentive to reposition itself than those that do not. When companies use telecommunications in this way, they often move into someone else's territory or find a profitable vacuum to fill. Reuters, Dun and Bradstreet, and GEISCO are clear winners here.

So too, is the National Securities Dealers Association (NASDAQ), which has operated as a computer-based trading system since 1981. In 1982 it introduced its National Market System (NMS), which entirely dispenses with the idea of a physical trading floor. Four hundred and fifty brokers cover 4,600 stocks, and the prices and volumes for deals made through the system are reported within 90 seconds. There are 8,000 terminals outside the United States and stocks of more than 300 foreign companies are traded. Japanese firms have often preferred to be quoted on NASDAQ rather than on the American Stock Exchange. NASDAQ's growth has been far higher than that of the New York and American exchanges, which did not grasp the opportunity of telecommunications as quickly, even though the importance of doing so was as obvious to them as to NASDAQ.

One question which the example of NASDAQ raises is "Why so often in the electronic marketplace can some organizations move so quickly to fill a vacuum?" A corollary question, more relevant to Dun and Bradstreet, is "How do some spot the right opportunity and succeed in turning it into reality and revenue while others see the opportunity and somehow make the wrong choice of what to do?" A third question, very applicable to Citibank, is, why can one firm do so well in exploiting the market opportunities opened up by electronic delivery even though the quality of its technical staff and many of its specific decisions were relatively poor?"

The features common to the firms that have successfully repositioned themselves throughout the electronic marketplace point to some answers to these questions:

- The winners have thought through very clearly what they want to be as a company.

- They did not get trapped by the technology and hope that it would in itself give them an edge.

- They focus on the infrastructure, the architecture, not the applications.

PIGGYBACKING AND NETWORK INTERCONNECTION

All these points are especially important for the last and, in some ways, the most far-reaching use of telecommunications for market innovation: the combination of services traditionally provided separately and (often by separate industries) at a single workstation. The network infrastructure becomes a multiservice facility. Because the infrastructure is already in place, new types of traffic can be added at relatively low incremental cost.

Several industries have lost control of their own distribution system as a result of piggybacking. Travel agents now face third-party competition from credit card firms, electronic publishers, and, potentially, retailers, all of which hope to bypass the travel agent through corporate travel departments, self-ticketing machines, and self-reservation systems. Banks face threats from just about every industry where telecommunications is a central aspect of customer delivery or internal operations. "Banking" becomes the payment component of other services: Check cashing in the supermarket, credit card transactions, or moving cash balances across the firm. Many banks have tended to have been not imaginative enough to understand the dynamics of the marketplace or not close enough to the customer to be able to anticipate the opportunities and threats of telecommunications.

Who is our competitor now? Five years from now? Who will control the new distribution system for customer delivery? What is the telecommunications infrastructure we need for either or both competitive advantage or defensive necessity?

These are questions to ask now. Once the marketplace has made the answers clear, it is too late for many firms to do much about the situation.

Controlling the Delivery Base:
Retailers versus Banks

EFT/POS is a well-recognized target of business opportunity that relies on interconnecting retailers' and banks' networks. It has taken a long time to evolve. Some day the retail store's point-of-sale terminal will handle many payment services that are traditionally "banking" and the debit card will complement the credit card. Both types will be "smart" cards, which can contain transaction information and provide extra security features over the standard card with a magnetic strip.

Logically, the banks should control the progress of EFT/POS. In practice, it has been retailers (and oil companies) that set the pace. They have been able to move far faster than the banks if they already have a better telecommunications infrastructure. A typical banking network can handle only 50–100 transactions per second. A customer using an automated teller machine usually makes only one or two transactions and does not mind a few seconds delay in response from the machine.

When shoppers are waiting at a busy checkout counter, delays in authorizing payments or transferring funds from their bank account are not acceptable. A retail point-of-sale network needs to handle 500–1,000 transactions per second, or even more. This has meant that it is far easier for leading retail stores to make EFT/POS work efficiently than for banks to do so.

Several of them have already succeeded in taking the initiative away from banks. Giant Food installed minicomputers in its 135 stores, linked by a private telecommunications network. It purchased a software package for telecommunications that had previously been used only by financial institutions. This software allowed Giant to operate its own ATMs and point-of-sales system over the same network. It was in a position to convert its existing check-cashing courtesy card to debit cards and link into the banking system's automated clearing houses (ACH) to process customer checks. It does not have to pay fees to banks for EFT/POS.

Publix, the Florida chain of supermarkets, went much further. It accounts for 25 percent of the food shopping market in Florida and operates close to 300 branches. In 1981 it began to implement a simple idea: replace check cashing for customers, which cost it 40 cents a check to process. Of this, 7 cents went to the bank. Publix planned to charge the bank instead. That, as one banker commented, amounts to role reversal. The large banks planned to install a proprietary network of their own and squeeze out the smaller banks. They waited too long. Publix gave access to its network to savings and loans, credit unions, and small banks. It was able to operate well below their costs. The average fees for the 10 largest shared ATM networks were 53 cents for a withdrawal; Publix charged 40 cents. Charges for deposits were 83 cents as against 60 cents with Publix. This is a measure of the economies of scale and efficiency that can be obtained in telecommunications.

The banks retaliated by installing the Honor system, a year later. This linked 1,500 machines, 500 of which were placed in grocery stores. Publix still set the pace, though. The banks had to sign a deal with it to allow Honor cardholders access to Publix ATMs.

Publix had become a new type of financial utility. It had potential income from service to millions of bank card holders, and from 5,000 financial institutions. It arranged for Citibank cardholders to use its system; Florida gets 40 million tourists a year, many from New York. A banker commented that this was admitting the Trojan Horse.

By the end of 1984 Publix had moved to full EFT/POS. Its Presto! system "lets you pay for your groceries with your ATM card." At first, Publix lost about $200,000 a month on its move into banking. It broke even in late 1984 and in mid-1985 was processing a million transactions a month, amounting to $50 million in withdrawals of funds and $15 million in deposits. It gave the state's financial institutions representation in prime areas; its customer count was around 20,000 a week.

There were obvious benefits for Publix. They included quicker traffic through the checkout counters, income from financial institutions, and many marketing and promotion opportunities as being up-to-date and giving customers convenience. It found that customer volumes increased 30 percent and the number of checks decreased 20 percent in stores that installed the system.

The Customer Has a Network, Too

All that the many large multinational firms with large international networks to coordinate their international operations need in order to find ways of profiting by piggybacking someone else's traffic onto their facility is a focused business concept. Texaco had that when it decided to buy half of a London foreign exchange brokerage and Ford when it tried, without success, to become a member of SWIFT. They both realized that they had better networks than the financial institution they dealt with and paid to handle funds transfers and foreign exchange dealing.

In 1982 Texaco had around $3 billion of liquid assets that it could invest in the Eurocurrency market, bonds, and short-term instruments. As an international firm it made frequent foreign exchange transactions, using a bank. The bank's dealers used broking networks to make the trades. It bought part of the London brokerage so that it would get the benefits of speed and directness in making transition, anonymity (to avoid affecting the going rates), up-to-the-minute information of shifts in the market, and the best possible deal (the 0.125 percent difference in rates across the money markets is small but adds up quickly on daily multimillion dollar deals). A CEO of another firm welcomed Texaco's initiative: "Texaco's now got itself a network like any major international bank. . . . We've been at the mercy of the banks too long." For Texaco, the total number of broking transactions a day would be small in comparison to the engineering data, electronic mail, financial control information, customer orders, etc., flowing across its network. Why pass important aspects of business over to intermediaries when they can be better handled directly?

Intercompany Communications, the New Business Environment

Business in the late 1980s is already moving very quickly toward interorganizational electronic transactions. Network interconnection between firms will soon become a norm not an innovation. If a company already has a strong communications capability, it will find many ways of exploiting it by linking its network to someone else's. The most obvious trends, which have already been described in this and the preceding chapters are: linking suppliers and customers, electronic data interchange to eliminate delays and documents, electronic trade management, credit cards as the base for capturing loyalty (and customer data), and electronic funds transfers at point of sale.

Obviously, many large organizations can afford to wait until the standards for interorganizational electronic business are fully in place and then join the bandwagon—or can they? General Motors in mid-1987 had eliminated all hard copy paper scheduling in interactions with its 1,600 production suppliers to its Canadian Chevrolet-Pontiac division. Ninety-nine percent of the suppliers had some form of electronic links in place. They have had to; not being able to link electronically to GM means not being qualified to be a supplier.

One in 4 shipments entering U.S. shipping ports from abroad is now documented electronically; the U.S. Customs Service expects this to move to 1 in 2 well before the end of 1988. Companies like Philips, the Dutch multinational, have been actively working on international and interorganizational EDI since 1985. EFT/POS is a reality in banking, retailing, and petrochemical firms. Over 25 percent of all grocery industry transactions between firms are now handled electronically, as are 80-90 percent of railroad waybills. General Motors, a bellwether for interorganizational electronic business, started implementing in early 1987 its plan to link 8 banks to 8 of its divisions. This will allow it to pay over 300 suppliers electronically.

Etc., etc. Will it be practical for any large firm in any industry to compete by 1990 without joining what is already an accelerating trend and soon to be an established practice?

Firms now have to define their technical architecture for the telecommunications highway system not just to meet their own internal needs but to be able to do business with a wide range of other firms. In many instances, the new markets of the 1990s are being created out of the combination of old ones via telecommunications. Electronic funds transfer at point of sale links banking and retailing, and banking and gas station sales. EDI saves Supr-Valu, the largest grocery wholesaler in the U.S. $1.30 on each of its daily purchase orders, cuts a day in lead time for delivery, worth $2 million, and allows 24-hour turnaround on document processing and return with over 700 other firms. Here is the new cooperative electronic marketplace benefiting customer and supplier.

GM has extended its requirement that suppliers eliminate paper and substitute electronic documents to banks; they are just another supplier. It asked eight banks to link to eight of its divisions to pay 300 suppliers electronically. One other bank, invited to join the venture, refused. It later tried to get back in, recognizing that here is a new electronic market that can only grow and grow.

The message is clear. A firm's communications capabilities determine its ability to share in these new markets. More and more business growth in the next decade will come from being able to link one's own network to someone else's.

TRYING TO CATCH UP TO THE NEW LEADERS

The examples in this chapter cover a range of industries and applications. They show the ways in which telecommunications opens opportunities for market innovation. Some of them change entire industries. Others are less far-reaching but change the dynamics of competition.

Some of them are little more than gambles on the future; the casualties in electronic publishing, point of sale, integrated financial services, and computer-integrated manufacturing have been and will be immense. There is no reason for any senior executive to stake the firm on any one of them. Leading too early can eat up money. Following too late, though, can make spending money irrelevant.

When do we lead and when do we follow? This question will occupy managers' attention for some time. Figure 5–1 summarizes the issues for deciding. It shows two extremes that define the technical base needed to support a particular business strategy: *fast installation* and *integrated operations*.

Figure 5–1.   Gaining Edge versus Ability To Catch Up.

	TECHNOLOGY BASE NEEDED TO MAKE A MAJOR MARKET OR PRODUCT INITIATIVE	
Type of Competitive Initiative	*Fast Installation Base*	*Integrated Operations Base*
PREEMPTIVE STRIKE	**A** • Uses external telecommunications service; can add capacity quickly • Software and processing can be purchased and installed at short notice • Competitors *can* respond and imitate or reverse engineer the innovation • The cost to catch up may be quite high • The original innovator is likely to gain the sustainable advantage of occupancy	**B** • Involves long lead times in building telecommunications and processing base • Customer acceptance forces competitors to respond, but they can only offer surrogates • Catch-up is measured in years • The innovator can lock in a whole market
NEW PRODUCTS OR SERVICES	**C** • No long-term technical or economic edge for the innovator • The leader takes the risks and in effect does the market research for the followers to exploit • The leader may be able to skim the market and gain advantages from being seen as the innovator	**D** • The innovator can exploit piggy-backing and add services more quickly and cheaply than the followers • The ownership of a reliable, high-volume electronic delivery base is a leverageable source of differentiation

The Questions for Top Management:
- When do we lead, when do we follow?
- What electronic delivery base do we need to create a competitive advantage? For defensive necessity?

In the first case, the communications capacity and links can be acquired quickly, particularly by using public networks. The software and hardware can be bought "off the shelf" and do not need much custom tailoring or investment in development. Capacity can be added in increments.

In the second instance, there is a far more complex interdependence among hardware, software for data management, transaction processing, communications switches, and communications links and protocols. Obviously, if a competitor's initiative is built on this integrated operations base, catch-up will be far harder than if it can be imitated by buying the facilities for fast installation.

A preemptive strike is only a little bit preemptive in the later instance (cell A in Figure 5–1). Competitors can respond, though they may have to invest heavily to do so. The originator of the move has to continue to build on it to maintain its lead and may be able to exploit the advantage of occupancy. Electronic cash management is one example of this.

Merrill Lynch's CMA, American Airlines AAdvantage program and reservation systems, and American Hospital Supply Corporation's distribution system each built on an integrated operations base. Competitors had to respond but could only do so with surrogates. Catch-up has taken years. This is cell B.

THE FUTURE: ARE THERE ANY MORE LIKELY BREAKAWAYS AND PREEMPTIVE STRIKES?

Were American Hospital Supply, Reuters, Citibank, Merrill Lynch, and American Airlines just special cases? Luck? Historical accidents? Can telecommunications really create sustainable competitive advantage? Is all this just infohype?

These questions were raised at the beginning of this chapter, with the promise to show that the old stories still have messages for firms' future options and risks, whether viewed in terms of greed or of fear, the two great creators of management awareness and action.

There will be more breakaways. It will not be easy to guess which firm in an industry will make the moves and succeed, but some targets of opportunity are clear. Look for the bank that uses its electronic delivery and cross-selling base to scythe into the largely complacent insurance industry whose cost base is at least 35 percent too high and that relies on regulation as its protection against outsiders. Watch for the oil company that uses smart cards (cards that can store information on the customer, past transactions, authorizations, etc.) to create brand loyalty for an undifferentiated product, through frequent driver programs—"We update your driver bonus record every time you use the card to buy gas." "Salepersons, stop at the Texexxomobshell sign—200 gallons and you get. . . ." "Drive safely with our card. It's loaded with preauthorized financing for up to $800 of car repairs if your car breaks down."

Keep an eye open for and then rush to travel on the airline that uses a smart card to entirely eliminate the ticket that is not really a ticket but a contract. You will be able to bypass the check-in desk and use your electronic boarding pass embedded in the card, which is also far more secure than that piece of paper. That airline will also probably persuade your elderly parents or grandparents to travel with them, too, because their card stores medical information and contact names.

Check which retailer uses electronic data interchange to shorten its order-to-deliver leadtime to a day and somehow always has the right items in stock to be able to meet demand but never carries buffer inventories; or the car manufacturer that at last matches Japanese quality via computer-integrated manufacturing and cuts anywhere up to $500 off its cost-per-car via electronic document interchange and reduces delays in handling dealer inquiries from weeks to a day via dealer automation.

Every one of these examples will occur within the next two years and create one big winner—if it can sustain its lead through the seven-year window of opportunity created by the advantage of occupancy and lead time. Perhaps some of them may happen before this book is published.

III THE SENIOR MANAGEMENT PLANNING AGENDA

Opportunities are not easily turned into actualities. The management process in most large organizations is an obstacle to doing this. It has been marked by three main features, all of which reflect the problems in the 1970s of getting data processing under control: delegation to a technical elite, inappropriate business justification, and the ethos of automation.

Few senior managers have known how to or wanted to play an active role in decisions about the deployment of computers and communications. They have relied on technical units that have until recently been isolated from the wider organization and that have often lacked credibility with business units.

The rapid growth of microcomputers, which took away much of the data processing unit's mystique and control over the technology, also reinforced its reputation for being bureaucratic and unresponsive. It is only in perhaps 1 in 10 large corporations that senior management has created a climate where the information systems function is seen as part of the dialogue throughout the firm about the basics of business. Instead, in most companies it is boxed into being a separate specialist group that occasionally makes proposals that have business implications. Management makes go/no go decisions in response to them and then delegates responsibility for their implementation.

The problem of delegation is probably the biggest single obstacle to progress in creating an effective way to mesh the business and technical sides of planning for telecommunications. Certainly, the most frequently expressed frustration of the telecommunications managers who understand the topics that were discussed in Part II is simply "How do we establish a dialogue with top management?"

A major cause of the frustration is the naivete and inappropriate business justifications that still dominate discussion of telecommunications: cost-displacement, short-term planning horizons, and full allocation of costs. That is no way to justify a capital investment with long lead times or an infrastructure that is intended to create opportunities for adding applications that improve revenues or costs.

It is impossible to cost-justify business innovation. Of course costs are an essential concern, but they have to be viewed in the context of the economics of doing business over the middle to long term and not in terms of direct costs savings and accounting mechanisms for allocating expenditures to user budgets.

Even though the entire thrust in the use of the information technologies has shifted from automation of clerical procedures to support of business functions, old images still dominate managers' thinking about their planning and use. They focus too much on the visible parts of the technology, especially the equipment, and on single applications. Their vocabulary centers around issues of staff savings, cost savings, and the impact on operations.

That is hardly unreasonable. Managers remember the traumas and fiascos every large company experienced in automating its clerical processes in the 1970s—the late projects, the ones that had to be written off, the cost overruns, the constant promises of huge benefits that somehow only resulted in more costs and huger promises. They have had good reasons to take a very hard-nosed view of the computing and telecommunications investment.

They have dropped their guard over the past few years. The personal computer changed their expectations and opened up their thinking. Office technology, especially word processing and electronic mail, offered potentially high improvements in productivity. "Expert" systems added new promises.

The promises have also created a backlash. Office technology has not worked its magic. Personal computers are only as effective as the people who use them and the usefulness of the tasks they are applied to. Computer expenditures are once again threatening to get out of control. Managers are demanding more accountability from their technical planners. The ethos of automation is coming back. It is almost as if there is a pendulum that swings the field of computers and communications from innovation to discipline to innovation and back again, from loose promises to tight control and away again.

This means cycles of boom and bust for telecommunications planning, or rather cycles of naive acceptance of the promises and then naive insistence on control. This is not a sensible way to balance innovation and discipline.

REDEFINING THE MANAGEMENT PROCESS

Delegation, inappropriate business justification, and the ethos of automation have to be replaced, and it has to be senior management that stimulates the replacement. The planning process can easily be changed—if the people at the top want it to be. Delegation is not a strategy.

There are four main requirements for an effective process:

1. *Stop extrapolating from the status quo* in forecasting and competitive analysis; recognize instead the need to scan nonindustry and nontraditional competition, both to recognize lessons and general messages and to anticipate (or initiate) intrusions on others' traditional territory.
2. *Avoid automating the status quo* and replace the mindset of automation with practical imagination, vision, and realism.
3. *Replace delegation* with management policy choices that highlight the trade-offs between business and technical options.
4. *Make the business case* in terms of business value first and then assess costs; distinguish between radical moves that need a business act of faith and operational ones where cost is the central issue.

The next three chapters present ways of redefining the management process. The methods are fairly simple. They are a synthesis and sometimes a systemization of approaches the successful innovators in the electronic marketplace have used informally.

Common mistakes senior managers make when they recognize the importance of telecommunications are to look for simple, almost magical concepts or to get involved in the complex details of technical planning. Neither extreme works. The simplistic one has done a lot of damage in the past few years. It includes the various overselling of sound concepts, especially office technology and artificial intelligence, where management is told that a few basic ideas somehow translate into fairly immediate and dramatic gains. The strategic message is often a valid one, but there is no link from strategy to the details of design, implementation, and use. Instead, there is lots of handwaving and the issue of lead time is obscured or even denied.

At the other extreme, the details can drive out any sense of strategy. Management gets bogged down in reviewing elaborate network plans, volumes of cost-benefit and traffic analyses, and scholastic debates about important but obscure technical issues. The business priorities become hidden under the operational details.

It can be hard to find a midpoint between these extremes. It is essential to begin at a level of abstraction high enough to highlight the key business criteria and technical design issues. It is vital to carry the analysis down to the point where the practical and tough details that dominate telecommunications implementation and operations are recognized and addressed. At that point, the process can move away from needing the senior manager's attention into the domain of the professional telecommunications specialist.

The process outlined in the next three chapters moves down the levels. It begins with building the business vision. Without that, there is little worth in taking up the senior manager's time and attention.

6 BUILDING THE VISION

Vision is a photograph of the future: something concrete that can be easily understood. The concreteness is essential, since telecommunications involves substantial uncertainties and ambiguities. It is very hard for people to have a clear idea of just how new electronic delivery mechanisms, computer-aided manufacturing, a whole new style of customer service, document interchange, videoconferencing, office technology, and so on will change the daily life of the firm. What will it be like to work here? What will I do on a sales call? What new types of decisions will I have to make? What do we tell our customers? Who is in charge of making the transition to a new style of doing business? What new skills do my people need?

The uncertainties are technical, economic, and cultural. It is difficult to think concretely about something that has never existed, especially when there is no established tradition for looking at telecommunications in business terms. There has to be a formal process to build a shared vision. It has three main goals:

1. *Shift the focus and terms of debate* for telecommunications from technology to business, and from cost to benefit.
2. *Provide a forum for sharing* views and building momentum and consensus and bring business people directly into what has up to now been a technical debate.
3. *Send the message* across, down, and up the organization.

The process can be handled in one of four ways, summarized in Figure 6–1. It may begin at the top, if strategic goals are already well defined and the role of telecommunications in supporting them needs to be clarified. It can begin at middle levels of the firm and then move up and outwards across it, as when telecommunications has been

Figure 6–1. How to Get Moving on Building the Business Vision for Telecommunications.

	DIRECTION OF MOVEMENT	
	Top-Down	*Middle-Up*
	• More suitable when business goals are clear and telecommunications already recognized as a significant potential contributor to meeting them • Work with the business leadership to set priorities and use results to stimulate interest at middle levels and communicate the business message for telecommunications	• Needed when the opportunities of telecommunications are not widely recognized at the top • Open up a dialogue at middle levels of the firm and use results to draw top managers' attention to their need to set direction
CATALYST FOR PROCESS		
Internal Senior planner and/or telecommunications manager credible with top management and business units	• Get directly to work on priorities • Clarify the business message to guide planning • Focus on what the firm can do and how it will benefit • Next step is often education and awareness program aimed at middle to senior business managers	• Build a consensus on opportunities • Stimulate ideas • Focus on overlooked opportunities and benefits from taking a longer-term strategic business view rather than an operational view of telecommunications
External Outside consultant or academic "guru" with broad knowledge of business and competitive links between telecommunications and business	• Begin a process of education and analysis to highlight strategic competitive issues • Focus on the competitive context • Next steps likely to use internal staff to work on priorities and payoffs	• Stimulate ideas • Use outsider to ask questions, pull together opinions, and identify where the firm may be overlooking opportunities • Focus on using education to capture interest and build momentum

treated mainly in terms of a technical utility and a broader recognition of the business opportunity needs to be fostered.

The process needs a catalyst. This may be an outsider who has the advantage of a breadth of perspective across industries and the right to ask the "stupid" questions and get the explanations of the "obvious" that can help break open mindsets and move people's thinking away from extrapolating from the status quo. It may be an insider who brings knowledge of the organization, credibility, and influence.

IDEAS INTO ACTION

A typical sequence for building the business vision is that used by one of the top international banks. The process had three stages.

1. *Define the "outside" vision.* Look at the industry and marketplace in order to get a sense of strategic trends and options; the output of this was a brief vision paper.
2. *Build the internal vision.* Use the vision paper to stimulate discussion and to move the issue up to the executive committee. The goal was to get a formal policy commitment.
3. *Turn the vision into a plan.* Use that commitment to create the necessary plan, make the business case, and move on to action.

The first two steps took six months and the third just under that.

The process used an external catalyst plus internal "thinkers" who had access to the "doers." The thinkers were staff people who were credible with the senior managers who could turn ideas into action. They produced a 20-page vision paper with the following headings:

- What is happening in the marketplace
- Where we stand with telecommunications
- Key components of telecommunications
- Our vision
- Strategic technical decisions
- Strategic organizational decisions
- Next steps

The paper was circulated among the key doers and frontliners, the people who generate the revenues, the bank's account officers, marketing managers, and branch heads. If they did not see the new thrust as exciting, beneficial, and practical, top management commitment would soon decay into middle management ignorance, indifference, and even opposition.

The VP in charge of developing this paper produced a one-page summary of it:

> We will remain a bank and not become a financial supermarket.
>
> Neither time nor geography will be a barrier to delivery of our main banking products.
>
> We will match the best level of customer service available in the industry in all major banking areas.
>
> We will build an international global communications network.
>
> We will not be preempted in our main markets, in terms of products and service by any bank or near-bank.
>
> We will start now to educate all our people to help them adapt to an era when we will deliver services almost entirely electronically.

The bank's executive committee agreed, as a result, to spend between $15 million and $40 million, depending on the detailed recommendations from a new planning group, on building a global network. It has become the first non-U.S. bank to do this, directly as the result of the vision exercise. The process the bank used to get ready for action is generalizable to any large firm.

Building the External Vision.   Bring together technical thinkers, who have access to doers, to establish a focus on the competitive context for telecommunications. The output is the brief vision paper.

Building the Internal Vision.   Get the vision paper to management doers and to the frontliners, the people who make money for the firm. The paper is a stimulus for thinking, a "straw man," and a force for change: if it is effective in its choice of examples and reading of the firm's competitive pressures and opportunities, it is unlikely to leave managers with the feeling that the firm can afford to treat telecommunications as an operational issue. The output from this second phase is an extended version of the original paper that makes more specific broad recommendations. The recommendations need to be specific about the business aims, but not about the technical and financial details. That requires a far more formal planning process.

Senior Management Briefing.   The vision process begins with divergent thinking—opening up possibilities and stimulating practical imagination. It has to end with convergence, not on detailed plans, but on principles for action. In almost all large firms, this means a briefing session for senior management which is partially educational. It will usually correspond to the sequence of the vision paper. The final session of the senior management briefing, which usually takes a day, has to address the policy agenda and may take up half the time for the entire effort.

CHOOSING AN APPROACH

There is a lot of fog in many firms about the extent to which telecommunications will affect their future and need their attention. Sometimes the fog simply reflects lack of information. That is easily remedied by education. Sometimes, it is due to lack of time and attention; managers recognize that something is happening around telecommunications but have not thought through what that means for the business as a whole. Sometimes there is clear air at the top of the company but fog down below; senior management has looked ahead and understands the importance of telecommunications for its future but middle managers have no awareness of this. Sometimes the fog is in the corporate office; executives do not see that telecommunications and business strategy go together for their firm or why they must move fast.

The choice of approach to building the business vision for telecommunications partly depends on where there is fog and where there is clarity.

If the fog about telecommunications is at the top, it has to be cleared at the top. Senior managers must be briefed so that they understand and act on the policy issues. They must understand that the technology is not the main issue. They have to set the scene for action.

Whether to rely on an inside or outside catalyst again depends on where there is fog. If senior management is unfamiliar with the competitive issues that were the topic of Part II of this book, they may well have to go outside to bring in a broader perspective than they and their people have. First, though, they need to check if this is a situation where their own telecommunications group does in fact understand the business issues and wants to bring them to management's attention but feels blocked from doing so. Quite often, it is they who bring the outside catalyst because technical people are not treated with the same respect as academic gurus or noted consultants.

Until they are, most firms will tend to rely on outsiders to begin the process of rethinking telecommunications. When they do, they ought also to ask, "Do our own people have the knowledge and skills to have done this for us?" If the answer is yes, then they need to ask why it was that the company had to bring in an outside expert to tell management something the firm could have learned from its own resources. The answer is usually, "It was too foggy for us to see clearly."

A CAUTIONARY NOTE: VISIONS
ARE NOT CONTAGIOUS

It can be very hard in any organization to change mindsets. It may be even harder to stop people extrapolating from the status quo. Ironically, it is often the very best firms that are most locked into old views and established ideas. Their sense of cohesion leads them if not to smugness then to a tendency to resist any challenge to the assumptions that have brought them success. Their industry leadership makes it hard for them to imagine that anyone outside the industry can be a threat. Their past achievements make them overconfident in forecasting.

Forecasting extends the status quo. The vision process challenges it. It is often the companies that are holding on—not quite failing but not succeeding as well as they expect—that are willing to examine basic axioms and rethink conventional wisdom. Telecommunications for business strategy demands that. It does not happen naturally and there are many forces in the organization that can blur the vision or even dismiss it as fantasy. Here are just a few examples that all highlight the essential need for senior business managers to lead, to turn awareness into action, to show their own commitment, and then to work hard to push commitment down.

Lawson-Hines (L-H) is one of Europe's top automotive firms. In the late 1970s it had been a basket case with high costs, labor disputes, and declining quality levels. A new CEO and management team had pulled it back from the brink. The CEO saw information technology as an opportunity but admitted that he knew little about it. He authorized a review of the firm's activities and opportunities.

The result was close to a fiasco. The outside consulting firm brought in found itself in what one of its partners described as "a meeting of theologians in Beirut—bring your own machine gun."

Divisional data processing units saw themselves as being on the line. They had a high degree of autonomy and authority and were suspicious of any initiative that would strengthen the corporate information systems function. Corporate IS, which played only an advisory role and was essentially ignored by the divisions, obviously took the opposite view. Administrative services, which runs voice facilities and premises, saw a clear need for a central planning unit; it nobly volunteered to take on the extra load and budget.

L-H is an organization with plenty of business vision in its operating units, but none of it is shared across them. The truck division has an integrated electronic cash management and payments system, which saves it millions of dollars a year, one car division has a comprehensive dealer network and has moved aggressively to exploit office technology, reducing administrative and personnel costs as a direct result. Corporate finance has been very successful in its use of personal computers.

There is no shared vision for telecommunications but lots of self-protective fiefdoms. The units are all busy reinventing the wheel. They have a strong investment in the business and organizational status quo and yet at the same time see a need for L-H to do "something" about companywide telecommunications. The topic comes up again and again at senior management meetings and again and again nothing happens. The division managers are the driving force in L-H's diversified operations and must remain so. Only if they as a group see telecommunications as a shared priority will change occur.

There is no internal catalyst. There is no "hook," no single business theme that can grab everyone's attention and point the direction toward cooperation. Perhaps the hook is computer-integrated manufacturing, but there is no one inside L-H to lead the dialogue and champion the integrated approach CIM needs and no one outside who gets listened to. How can anyone get the message to top management? Where are the strategic thinkers with access to the doers? L-H is not just a "decentralized" but a diffuse organization and it is unlikely that anything will happen until L-H finds out from its competitors what it needed to start thinking about years ago.

Vision is about opportunity. Its enemy is fog. Heller is a major international insurance firm in the United Kingdom. It is part of the famous Lloyds' market, which is under increasing pressure as international insurance follows the same historical trends as international banking. Telecommunications will then be a major factor in shifting the historical center of gravity of the market away from London. Already, 55 percent of world business originates in New York and almost 20 percent in Tokyo. London's share is well under 10 percent.

The business vision for any international player has to include telecommunications. Heller's head of information systems carried out an imaginative and practical vision study, which as he commented "came up with the self-evident":

> We will run New York out of London and vice versa, in effect giving our U.S. brokers direct access to our back-office. . . . We will ensure as close to perfect ease of operation and speed of communication across the world as is feasible.
>
> We will create ahead of our main competition new product opportunities by linking insurance payments and movement of funds, documentary credits, etc., with banks and customs services. . . .

Heller's CEO rejected the vision study's conclusions, although many of his younger executives were strong supporters of its recommendations. Like many of his fellows in the Lloyds' club, he is firmly convinced that Lloyds' will remain the dominant market. In reply to a consultant's report that stated "The Main Board (the team of 15 executives who make all key decisions) does not see information technology as a strategic issue," the CEO replied, "God help us the day we do."

That is fog or double vision. The vision of today will become tomorrow's common sense, as Heller will find out.

7 POLICY
The Bridge between Vision and Action

The end point of the business vision for a telecommunications strategy is a complex—and expensive—technical artifact composed of cables and boxes. In most firms the two do not connect. Technical staff become very suspicious of articulate "gurus" who talk in futuristic terms as if everything were easy to implement. They accuse them of not really understanding telecommunications. Business-oriented planners feel just as frustrated with "wireheads" who are unable or unwilling to look beyond the technology and talk as if everything were so hard to implement that nothing new should be tried except at their discretion. They accuse them of not really understanding telecommunications.

A bridge can be built between the two. It requires a recognition that business criteria should drive technical planning. They rarely do, though, because the technology quickly dominates discussion especially when it is largely new. The technical details are complex, and vague handwaving and slogans do obscure the practical difficulties. Technicians have to understand, though, that it is impossible to talk about network "optimization" without defining first what the firm is trying to optimize. Investing extra money to get "better" reliability, costs, or response time has to be justified in terms of some business payoff. Technical decisions are not absolutes but involve trade-offs with economic variables.

Moreover, business planning has to point directly to technical options. The broad, top-level business idea has to be stated in terms that allow it to be related to the technology at a level of detail that is specific enough for a good technical specialist to say, "This means we must do X or we can choose A or B, with these trade-offs."

Figure 7–1 shows a framework for working from vision, if not to cables, then to the point where the technical planner can take over and the business planner can pass on responsibility, amicably and with confidence in each other.

Application Opportunities.   The articulated business vision identifies the main priorities for applying the technology. For instance, if one goal is to reduce the time taken to process customer transactions from days to hours, the application might be to link branch offices to the head office, provide facilities for transferring purchase documents electronically, and allow some customers to initiate transactions from their own terminals.

If the vision were to be the most cost-effective firm in the industry and to eliminate corporate bureaucracy, the application might include computer-integrated manufacturing and distribution, and the use of office technology to improve managerial and professional productivity.

Policy Choices.   Policy is crucial to manage the change that will ensue from application change in jobs, costs, planning procedures, and the degree of risk the firm may have to accept. It is by focusing on policy that senior managers break away from delegation as the strategy. They have to ask what decisions must be made at the top so that others have clear guidelines for planning.

Constraints.   Policy removes some constraints on planning and implementation. Others remain. They include government regulation, the availability of technology, and industry practice.

Strategic Design Variables.   There is no one "best" design. Trade-offs must be made among four main factors:

1. *Capability.* What services can the facility provide and what volumes can it handle?
2. *Flexibility.* Can the telecommunications resource handle today's needs cost-efficiently and be easily changed to meet tomorrow's?
3. *Quality of service.* How reliable is it?
4. *Pricing and costing.* What is the unit of cost from the user's perspective and what are the dynamics of cost over ranges of time and volumes?

Figure 7–1. Strategic Telecommunications Variables.

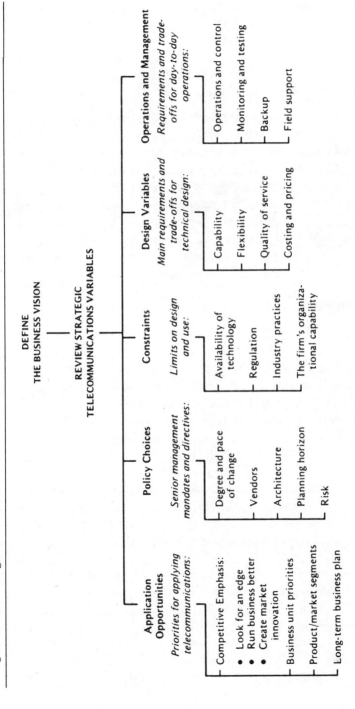

It is the balance between these factors with which technical designers have to deal. The cost of adding extra flexibility, for instance, means adopting an architecture that may have substantial overhead; it would be cheaper to use a facility that is optimized for today's traffic. For some application opportunities, reliability may be vital; for others it may be less critical than low cost of use.

Operation and Management Factors.   The final category of planning variables—operations and management factors—is of more concern to the telecommunications manager than to the senior executive, except when it has been ignored with the result that the system performs poorly. A useful analogy is with an electrical utility. Top management is concerned with strategy, funding of capital investments, pricing, regulation, and measures of service. When there is a sustained loss of service, from a brownout or failure of equipment, issues of maintenance, field support, backup, and troubleshooting move to the top of the list of management concerns.

Highlighting the Trade-offs

The framework in Figure 7–1 highlights trade-offs. Almost every one of them involves trading off a technical variable against a business one. Too much of the literature on telecommunications planning ignores nontechnical factors, but it is impossible to talk about "optimizing" a network without considering them.

Each of the five main categories shown in the figure can be broken down into levels of increasing detail, working from the top down, checking for consistency, highlighting areas where explicit policy decisions on trade-offs may be needed.

Security, for example, is an issue that will become more and more important as customers make transactions directly from their offices or homes. They need to be able to get access simply and cheaply. Standard personal identification numbers (PINs) and passwords do not provide adequate protection from error and abuse. Encryption, scrambling the message sent along the transmission link, adds security, as do hardware and software devices to authenticate the person making the transaction. They also add to the cost of the system. They may remove some design options such as the use of particular public data networks.

Is security a policy issue or one that can be left to the discretion of the network planner and thus reduced mainly to a trade-off between available technical options and cost? Too often, that question is not explicitly addressed. Very few banks and even fewer airlines, for instance, have a policy on security. They take an ad hoc approach that works only because most of their customers are unaware of just how low the levels of security are. Only when another group of teen-age "hackers" is arrested for tampering with top secret and "unbreakable" telecommunications networks do they wonder if their cash management system is secure. Automated teller machines are known as automatic theft machines among certain entrepreneurial segments of the population.

Senior managers in many banks recognize that security is impor-tant, but they trap their telecommunications planners in a double-bind by asking for both high security and low cost. That cannot be provided; it costs about $600 to ensure that a workstation is fully secure.

One value of the framework shown in Figure 7–1 is simply to high-light such issues. The telecommunications manager's job too often entails a Catch-22 like the one concerning security. Although it is not part of senior managers' responsibility or competence to resolve the technical issues relevant to designing a network, it helps immensely if the managers understand the main concerns and gear their policy decisions to eliminate double-binds.

RECOGNIZING OPPORTUNITIES

Part II of this book provided a framework for assessing opportunities: look for an edge in existing markets, run the business better, and find sources of market innovation. There are obviously other suitable approaches to classifying opportunities, such as the following, each of which has limitations:

Business Unit. The divisional or functional approach is an obvious starting point for evaluating using telecommunications but it ignores the opportunity to create a new service and find a market niche by integrating what are currently separate services.

Product or Market Segments. In some cases it makes most sense to focus on market-based groupings. For example, retail and corporate banking are very different in terms of customer needs, product requirements, and telecommunications traffic. It may be that all of the various business requirements can be accommodated on one tele-communications resource but more likely that several will be needed.

The only difficulty with the product or market approach is that it too easily becomes a self-confirming prediction. Companies already have too many separate, duplicated, and fragmented telecommunications facilities entirely because they have been built up in isolation to meet specific requirements at a given point in time.

One of the most frequent reasons decentralized firms often overlook ways in which they can exploit the business opportunities of telecommunications is that this requires a geographic rather than a functional or divisional perspective on the firm.

The Long-Range Business Plan.   This may be the obvious starting point if the plans already recognize the strategic business opportunities opened up by telecommunications. If not, then the plans are likely to reinforce moving along the current path, rather than cutting a new one.

Whatever classification is used to highlight application opportunities and clarify priorities, it must use business terms, not technical ones. "Electronic mail" or "videoconferencing" are not application opportunities, but the vehicles for realizing them. "Use telecommunications to improve coordination across time zones" is the opportunity.

It can be difficult to provide simple operational definitions of the end for which the technology is the means. That may be why there are so many rallying cries in the field of information technology that are intuitively appealing but that somehow never quite turn into the expected benefits. The most obvious of these are "Office Automation," "the Office of the Future," and "End-user Computing." "Artificial Intelligence" and "Expert Systems" are more recent ones.

They are not application opportunities, only signposts for looking at particular technologies in particular terms. Office automation, for instance, covers a range of technical building blocks; the original value of the term was its suggestion that firms should look at office work as a major opportunity for investing in information technology. It left the specifics rather vague, under the general heading of "improving productivity."

The operational definition of the application opportunity needs to be a whole sentence: a verb plus an object plus a qualifier. The verb defines the action to be taken, the object the target population, and the qualifier the business reason:

- Install workstations in key customers' offices in order to reduce the time taken to handle warranty claims.

- Provide all professionals with access to special databases to help them speed up and improve the quality of our analysis of competitors' marketing moves.

It is surprisingly rare that plans for telecommunications start from such statements.

POLICY CHOICES

Policy is the most neglected means for senior managers to play an active role in planning and for telecommunications managers to be able to break out of the frustrating straitjacket that binds them into cost displacement, short-term planning horizons, and technical tactics. Policy is an explicit set of mandates and directives. It defines bounds on planning and design and clarifies responsibilities and authority. It thus sets the criteria for the architecture and establishes the role of the architect.

Policy is partly a set of choices—"This is the way we will do things around here"—and partly self-imposed constraints—"We will only consider options that fall within these limits." In some areas no policy choice is needed, but that should be an explicit decision, not something that happens by default.

The Need for Policy: An Example

The issue of which vendors to use for transmission and switching has grown more complex because of deregulation and rapid advances in technology. Is a policy needed?

A firm could use any one of many suppliers and make its selection on the basis of cost and ability to meet particular technical specifications. But this is a time when the market is highly volatile and there is intense competition, with many small vendors entering the market (and occasionally being forced out of it). The largest ones are trying to establish control through their established customer and product base, marketing expertise, and research and development base.

There is still substantial uncertainty about standards and even more about the practicality of particular products and approaches. *Data Communications* magazine summarized the situation in mid-1985 in terms which are just as applicable in 1988: "how to avoid being buried by the new products avalanche," and quoted a consultant's opinion that the situation was "as wild as we've ever seen it. . . . There's too much confusion for the buyer. The products are nice, but they need improvement, they need to be easier to use and understand." (Bob Holland, in "Viewpoint," *Data Communications Extra*, Mid-May 1985, p. 11.) Holland also criticizes buyers' naivete, in terms that still remain valid and probably will continue to be so in 1990 in most firms.

> [Executives] are perfectly willing to delegate the responsibility of choosing to a lower-level technical person, without thinking about the overall strategy or agreement within the company. . . . Those people usually see their job as being limited to finding out which purchase would be the most technically sound. Their fact-finding, and finally, their decision-making may not be shared with the company's financial and telecommunications staff, and probably doesn't take into consideration . . . other corporate facilities outside their authority.

In addition, the "internal builders" too often disagree with each others' technical assessments. This is not a sensible approach, but the natural one when there is no policy directing efforts and attention. There are several policy options:

- Only vendors with a strong financial base will be considered. They must have a market share of at least 15 percent, a record of continued sales growth, and maintenance of profit margins.

- We will not consider any product which is not proven. Claims and specifications are irrelevant. It must have an established customer base, including at least 10 companies in the Fortune 100. It must have been in use by them for one year.

- We will select only products that are fully compatible with particular communications standards and computer hardware and operating systems.

- Whenever possible, we will use a single vendor for transmission and a single vendor for switching.

There is no one "right" policy, but there must be some policy. The choice reduces the range of options that can be considered at the detailed design level. That has the possible disadvantage of missing out on an exciting innovation or risking being locked into particular vendors or technologies. It has the corresponding advantage, though, of establishing key criteria, clarifying operating assumptions, and eliminating arguments later in the design process.

The policy not to have a policy should be an explicit choice and not one made by default, tradition, or accident.

The following list identifies the major policy issues that all firms have to address if they are to link vision to architecture, end to means:

- *The degree and pace of change* management is willing to accept
- *Vendor policy*
- *Architecture policy* and the range of services to be accommodated by it (not the specific design of it)
- *Planning horizon,* including funding and cost recovery
- *The level of risk,* technical and financial, that is acceptable

Other items can be added. For example, security may or may not be a policy issue. It certainly should be for any firm whose customers use its telecommunications network to make transactions that create direct transfers of funds or that require a high degree of privacy. One question for senior management to ask its business and technical planners is simply, "Is this list of policy items complete or is there another item we should add to the agenda because it requires a directive from the top?"

Degree of Change

How fast must a firm move for competitive reasons? How fast can it move without disrupting existing organization and operations? What degree of change will it accept, in terms of business activities, culture, shifts in jobs and skills? What resources does it need to commit to smoothing the process of change?

Many firms have overlooked the need to link the pace of technical change with that of organizational change. In financial services, for instance, the business strategy has often run way ahead of needed changes in skills in selling and the sales force's attitudes to making those changes. Citibank's and Merrill Lynch's aggressive account officers blocked rather than supported those firms' moves in the electronic marketplace.

One outstanding example of how hard it can be to keep all the components of the change process in step is Sears' ambitious strategy for capturing the broad financial relationship with the middle class. *Fortune* commented in October 1985 on the many problems Sears still faced in building its "struggling" financial empire. One cornerstone of this is Dean Witter Reynolds, its brokerage firm. Since 1982, Dean Witter's sales force has increased by 40 percent, bringing in 2,000 new brokers. It pulled in 154,000 new accounts from Sears stores in 1984 and closer to 250,000 in 1985.

But productivity dropped badly, from almost $100,000 in securities commissions in 1983, which was 86 percent of the average for brokers on the New York Stock Exchange, to $68,000, which is only 70 percent of the industry average. "When you bring so many people on stream at one time, it's harder to have the quality selection you'd get by expanding less rapidly," *Fortune* comments.

In early 1988, Sears looks a strong potential winner in the electronic marketplace. Its Discover card is well established (though not always used as often by consumers as their other credit cards). Sears management is clearly ready to stay the long course. Its subsidiary set up to coordinate telecommunications is probably among the five strongest units in U.S. business. But even so, Sears needs years before it will turn goals into achievement.

Sears's experience is a warning that the organizational change required by and implicit in the business strategy may not be at all smooth. There can be a conflict between the need to accelerate change to meet competitive pressures and demands to slow it down to meet cultural constraints. Which factor should set the pace?

The policy on the scale and rate of change that management wants has to be complemented by a decision to provide resources to make the policy effective. Increasingly, this means education. This is not the same as training, which generally follows implementation and focuses on teaching people how to use the new terminal and work procedures.

Education has to have more strategic goals, such as changing attitudes to the technology and its uses, building the skills needed to participate in the process of change, providing a forum for people to express and resolve their concerns, sharing information, and making sure staff understand the business vision and what it means for themselves and their jobs.

Education has to begin early and be made pervasive. It is one of the fastest growing parts of the information systems budget in most large firms, especially senior management education. Many of the companies that moved aggressively into the electronic marketplace in the early 1980s subsequently had to invest heavily in educating their sales forces. (See Chapter 10 on smoothing the path of change.)

Vendors

There are five main policy options, described in the following paragraphs.

A Primary Vendor for Both Telecommunications and Processing. Other vendors' products may be used but only if they are fully compatible with the primary supplier's standards. In practice, this means using IBM. This option raises the temperature of most firms' technical staff. Telecommunications specialists are not, in general, enthusiasts for IBM. Most information systems managers see IBM's architecture and operating systems environment as the de facto standard. Senior managers are sensitive to being locked into a single supplier.

The discussion can get heated, partly because of stereotypes about IBM. In the late 1960s, IBM and caution became synonymous in data processing. The competitors who took on IBM and dropped out of the marketplace included Xerox, General Electric, and Philips. Their failures left many data processing managers out on a limb and even out of their jobs. The cliché that no one is ever fired for going with IBM gained currency because of this.

In the late 1970s, IBM managed to misread almost every cue from the computer market. Now that its personal computer is the standard others have to follow (or create clones of it), it is easy to forget how late IBM was in accepting that microcomputers were more than a toy. It was sluggish too in minicomputers, office automation, and end-user computing.

Its telecommunications architecture, Systems Network Architecture (SNA), was also rigid and cumbersome. Its "Star" configuration (Figure 7–2) meant all communications had to go through a central mainframe computer. Minicomputer manufacturers, such as Digital Equipment, had designed their communications architectures to handle a nonhierarchical network. SNA was poorly suited to a world of office technology, where terminals ought to be able to talk to each

Figure 7–2.   Network Topologies (*T* = *Terminal*).

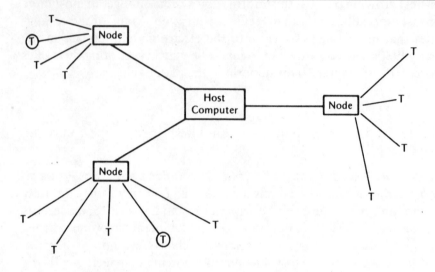

Star Topology

- Centralized host computer controls network
- Intermediate front-end processors ("Node")
- Terminals ("T ")
- In IBM's SNA, the Systems Services Control Point software in the host controls the communication between devices, such as the two circled terminals

other directly, "peer to peer," instead of being controlled by a central "host" computer.

In 1980, SNA was a weak contender for primacy in telecommunications. European governments, telecommunications agencies, and computer manufacturers were strong supporters of the effort to define the Open Systems Interconnection (OSI) reference model for telecommunications, which would provide a vendor-independent set of international standards. This was at least in part an overt move to protect their markets against IBM.

IBM woke up in the early 1980s and moved from laggard to a ferocious competitor in every segment. It also recognized the broader importance of telecommunications. It invested in Rolm, one of the top three makers of private branch exchange (PBX) communications switches and evolved SNA to the point where it became the de facto

Figure 7–2. continued.

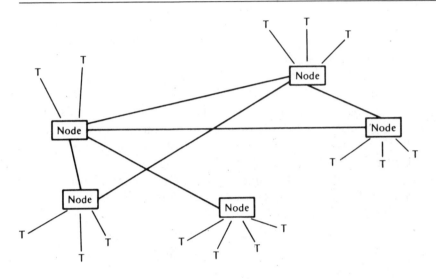

Distributed Topology

- No central network host: network is fully "distributed"
- No hierarchy, unlike Star
- Most public data networks designed on this basis
- Historical, technical, and conceptual incompatibility between SNA and X.25 standard reflects hierarchical versus nonhierarchical view of network management

standard for large firms' networks. OSI remained largely a paper model, with only parts of it being implemented in products.

In the mid-1980s, Digital Equipment Corporation (DEC) became a strong new threat to IBM's dominance, by attacking IBM's weaknesses in mid-size computers. A large part of DEC's success, though, came from abandoning its efforts to use its own proprietary telecommunications standards. It was DEC's decision to accept the need for full communications compatibility with IBM that made its existing advantages of the technical design and cost of its small to midsize computers into a powerful new competitor.

The history of IBM is important because it established many stereotypes about the company and data processing managers. There are many telecommunications specialists and even more academics who think of IBM as it was in 1980, not 1988. They see information systems managers who build their processing base around IBM's standards as neanderthals. They regard SNA as still centered around a mainframe view of computing, and push for OSI as the blueprint for action, especially now that IBM has conceded its future products will be based on OSI.

This perspective is partly emotional, too. There are many, many good technical specialists who simply dislike IBM, for what they see as its arrogance, the inadequacies of its products, and its historical emphasis on marketing at the expense of technical innovation. They see IBM's relatively sudden slump in late-1986 through 1987 as evidence of this. Senior managers need to recognize the complexity of the issues around IBM. To some extent, the policy issue around vendors comes down to the questions, "If not IBM, then who and why?"

It is important to get this issue out in the open and try to sort out what aspects of technical planners' opinions on IBM reflect taste rather than fact. This is as true for the proponents of IBM as of its opponents.

The main argument for a single supplier is the long-term strategic importance of integrating all aspects of the information technologies. The argument against it is that no one firm, including IBM, combines leadership in all areas. IBM, for instance, has been unable to make a strong move in the critical area of PBX, even after acquiring Rolm. AT&T, Northern Telecom, and Rolm together lead the market, but Rolm's products are still not fully meshed with IBM's.

A Primary Vendor for Telecommunications and Another Primary Vendor for Processing.   This used to mean AT&T plus IBM. Deregulation has altered this entirely. The breakup of the Bell system created immense disruptions in service.

AT&T, like IBM, has several competitive advantages: staying power and standards. The year 1985 saw the start of a large shake-up in the computer market. The same process was beginning in data communications. More and more announced new products were delayed, including IBM's local area network. Some firms that were highly dissatisfied with AT&T's marketing and service still argued that in the end they must stay with the company that has the financial clout to fund the huge investments in product development that will be essential for the rest of this century and beyond and that owns many of the key telecommunications transmission and switching standards. Some industry observers in 1987 openly doubted that MCI and Sprint, AT&T's maverick competition, could gain the 15 percent of the market needed to remain major players, even though several of their services were superior to AT&T's. Sprint had used its advertising budget well but had been buying market share at a loss that could never be sustained, close to a billion dollars.

A Formal Multivendor Plan.   Since no one company has solved all the problems of meeting large firms' telecommunications needs, let alone those involved in integrating communications and computing, there are strong arguments against using any primary vendor. Defining the criteria for using multiple vendors is much harder, though, than sticking with a few established leading suppliers. There is the obvious problem of incompatibility. According to consultant Bob Holland, a typical example is the "firm [that] had seven different kinds of office automation equipment, and the service carrier that had been contracted for had no way to link all the equipment together." ("Viewpoint," *Data Communications Extra,* mid-May 1985, p.12.) Managers reading the vendors' advertisements would never guess that, despite protocol conversion, IBM-compatibility, "integration," and standards, the pieces still do not fit together.

Any multivendor plan must be built around standards of compatibility. The policy may require using only vendors that can provide full compatibility with IBM's Systems Network Architecture (SNA) and support for its 3270 terminal protocol. Digital Equipment, Tandem, Amdahl, Northern Telecom, AT&T, Wang, and NCR are just a few of the communications and computer suppliers that have products that meet these criteria. In fact, it is possible though not probable for a firm to adopt IBM's architectures without having to use any of its products, combining Hitachi or Amdahl mainframes, Tandem switches, Compaq personal computers, and MCI's VNET network, for example.

Nontechnical Criteria for Selecting Vendors.   Two criteria of increasing importance are service and support. Telecommunications products are part of a utility; when there are problems, however minor, we want the lights back on—now. There are several major computer and communications suppliers whose reputation in the marketplace has been slipping over the past few years, not because of limitations in their products but because of poor service. As many of the products become commodities, support is the differentiator, not price.

The relationship between supplier and customer has to become one of partnership when the planning issues are complex. It is not a question of buying products off the shelf, but building mutual understanding and even joint development. IBM has always understood this. AT&T and others are having to learn it quickly. It is interesting to note the signal MCI was sending to its customers about its own role and the relationship it wants to build with them. At a seminar for its large corporate clients, it included a briefing titled, "What we expect from our customers."

Other criteria have already been mentioned: staying power, in terms of a broad customer base, strong balance sheet, funds for research and development, proven technology, and any other factor that can reduce risk.

These are not always appropriate policy requirements. There is no proven supplier of smart cards, for instance. Many of the most advanced products in specialized areas, which can substantially improve cost and efficiency of operations, come from firms newer and smaller than the established giants. This is why there has to be a policy decision.

Laissez Faire.   This is really not a policy but merely throwing up one's hands and giving up any effort to impose order on the chaos of communications supplies. "Do the best you can." Incompatibility, which is the opposite of integration, is the guaranteed outcome.

Even though it is not a policy, laissez faire is a surprisingly common practice, especially where local divisions have substantial autonomy over purchase decisions. Quite often, the corporate telecommunications group defines equipment standards, but these are not backed up by any degree of authority. In some instances, the central recommendations conflict with local needs, scale of operations, existing facilities, or, in the international sphere, with regulation and monopolies of the PTT, the official central communications agency.

Once the content of the policy is established, the central issue is *who* decides. If the issue of authority is ambiguous, the result is often political conflict between a "bureaucratic" central telecommunications staff unit and the field. Responsibility without authority has too often been the burden of the beleaguered telecommunications manager. The traffic on the corporate network is increasingly decentralized in terms of decision-making, development, and operations. It should not be pulled back into the center under a monolithic communications agency. Nor, though, can telecommunications be left to laissez faire planning and purchases.

The main issue for the policy agenda is to ensure consistency. If the policy establishes strong and tight criteria, such as reliance on a primary vendor, there has to be a central coordinator who oversees conformance with the policy and resolves the many exceptions that will need to be handled. If the policy is a loose one, there will be less need for defining complementary authority. The point is especially important in defining the criteria for the overall communications architecture, which is interrelated with that of the policy on vendors.

Policy on Architecture

Conceptually, it would make most sense to decide on the architecture and then the vendor policy. If there were a well-established set of universally adopted standards, that would be practical, because the choice of equipment and services would be relatively vendor-independent. Managers could shop around for the products that offered the best price or performance.

At present the potential architectures are far more determined by vendors than general standards. There is the promise of OSI, which has the active support of most computer makers and the more passive support of IBM. ISDN is the blueprint for the 1990s and onward. This is the computer equivalent of the international telephone system. Subscribers in the United States can dial direct to France, Spain, Japan, and almost every country in the world, even though the telephone systems have very different technical features. ISDN will establish the standards to interconnect all the applications of information technology in the same way.

But this will take time. Parts of ISDN will be in place within a few years. The bandwagon, moving leisurely but doggedly for years in Europe, was at a standstill in the U.S. until 1987. Now, even though many vendors and telecommunications managers are not quite sure what ISDN is or means, there is at last full agreement that ISDN is the end-point of the journey, even though the exact paths are unclear. (ISDN is often caricatured as Integrated Services Dream Network and Items Subscribers Don't Need.) The principles are already being included in the design of switching equipment, corporate and public networks, and terminals. The U.S. pilots in progress should bring results and plans for roll-out well before the end of 1988.

But even when the standards are agreed on, the detailed implementations differ, by country and by vendor. For example, X.25 is by far the most widely adopted basic transmission protocol and is used on almost every public data network. It is a true "standard." There are, however, many dialects of X.25 and one company's implementation may not be exactly the same as another's—they are incompatible and will not work effectively together.

Standards are emerging, and we can be sure that eventually ISDN will provide the integration of the many islands of information technology. That defines the goal. The starting point is where firms are now. That generally means multiple facilities and protocols. The main policy issues are:

- Do we want to commit to having a single overall architecture?
- If so, what are the timetable and major phases for rationalizing the existing incompatible components now in use?
- Do we need different architectures for particular applications, countries, or business units? Why?
- What range of services should be anticipated in the architecture? (The word is "anticipated" rather than "included," since the architecture is the blueprint for evolving an increasingly integrated resource over changes in time, technology, locations, users, and applications.)

The services include voice communications, document transfer, transaction processing, facsimile, conferencing, interconnection to other firms' networks, and access to remote information resources.

The business vision and application opportunities should obviously determine the policy for the architecture and the services it should be able to accommodate, not the reverse. The overall issue here is what level of integration is the firm's goal.

A financial service organization, like Citibank, Sears, or American Express, has to push for more rather than less integration, since its business vision rests on delivering a growing range of services through the same electronic delivery base to the same workstation. These do not all have to be provided through the same transmission links or be processed on particular hardware. One of the central components of the architecture (the most expensive) is the set of switches and data concentration equipment that converts one vendor's protocols to another and that interconnects networks, computers, and workstations which may be made by different manufacturers.

For financial service firms, the business vision implies an integrated corporate delivery system. Many manufacturing firms do not need this. They may, for example, want to operate special high-speed networks for computer-aided design, videoconferencing, electronic mail, or dealer transactions and inquiries. Each of these is very different in terms of length of messages, requirements for speed of transmission and response time, and trade-offs between cost and performance.

Business strategy guides the definition of the architecture. In many instances, senior management will just have to accept their technical specialists' conclusions and recommendations. They must, though, first be sure they get understandable answers to at least the following questions:

- Are our plans for integration or for maintaining separate facilities explicit and consistent with our business vision and main application opportunities?

- Do we have a target architecture as the end point of our investments for the next 5–10 years? What are the business and technical assumptions underlying our choice?

- Do you need us—senior management—to give you more direction, clarify business issues, or make policy decisions so that you can come back to us with a brief statement of the principles on which our architecture will be based?

The Planning Horizon

It is hard to imagine a business vision that aims to restrict innovation and that justifies a very short planning horizon. Generally, the planning horizon has to be 3–5 years, the lead time for intermediate moves in telecommunications. Although the architecture has to be designed to allow evolution over at least a 5–10 year horizon, there are too many business and technical uncertainties for firms to be able to lay out specific plans much beyond 3 years. That also means that there will be some level of central direction and funding.

Chapter 8 discusses how to make the business case for radical moves in telecommunications. Obviously, there has to be systematic business justification for the investment, even if the exact volumes, costs, and where relevant, revenues cannot be predicted.

Senior management too often creates a double-bind around telecommunications. It wants innovation and fairly immediate and tangible cost recovery. That simply will not work. Building or substantially expanding the communications infrastructure as part of an aggressive business strategy means spending money to get benefits later. The budgeting process for telecommunications in most firms means the planning horizon is too short to make innovation possible.

Risk

The last main area of policy where management has to make sure it has an explicit position, or gets the information from its business and technical planners needed to decide on one, is to some extent just a consolidation of the other policies: How much risk are we ready to accept?

Technological Risk.   Do our application opportunities, vendor policy, criteria for architecture, and planning horizon commit us to serious risk that the technology will not work as planned, be delayed, or involve unpredictable and substantial extra costs?

Financial Risk.   Can we limit the cost involved, if only approximately, such as not less than $X million and not more than $Y million over Z years?

Organizational Risk.   Can we handle this with our existing management and technical team? Can our people handle the strains and change it involves?

Business Risk.   How confident are we that customers will respond positively to our application objectives? How long will it be before we have a clear idea of the profitability of the innovation?

With telecommunications, high return almost naturally goes with high risk. Innovation involves some unavoidable gambles. If this new idea were easy to implement, someone else would be trying it, too. The edge has to come from doing it earlier, in most instances.

CONSTRAINTS

Most constraints on the design and use of the telecommunications resource are not directly controllable: government regulation or the availability of technology, for instance. Other factors such as organizational capabilities and industry practices are controllable in the long run, but act as distinct constraints on current and short-term activities. Recruitment and education can alleviate the first, and (sometimes) lobbying or negotiation the second.

Obviously it is important to decide which of these are indeed constraints, or at least are to be treated as such. It may be, for instance, that industry practices are more a tradition than a constraint. American managers in multinational firms will, however, be very surprised in many instances by the extent to which regulation is a binding constraint in most other countries. Constraints eliminate choices.

Availability of Technology

It is difficult to determine the extent to which availability of particular technologies is a constraint. For example, there are many opinions as to when fully digital public networks will be available in major countries. If the telecommunication planners assume that they will be in place as announced and that vendors of PBX will be able to meet their claims on performance, but the expected capabilities are not delivered on schedule, obviously the problems that result may be very damaging. If, on the other hand, the planners err on the conservative side and assume that they will be more constrained than is actually the case, they may, equally obviously, lose an opportunity to exploit the new technology.

Other areas of telecommunications technology in which experts disagree on availability include the pace at which integration of voice and data on the same facilities will become reliable and cost-effective, intrapremise communications links, and common standards. If experts disagree, how can managers make reliable forecasts? The issue is to decide what aspects of the technology will be treated as a constraint. Anything that is not a constraint is a potential design option and vice versa.

Predictability.   Are the trends clear enough, so that the planners can reliably count on a particular technology being available? For example, in the United States, satellites and fiber optics are ensuring that there will be plenty of capacity available. Bandwidth is not a constraint. International standards may still be.

Stability.   Whether predictable or not, is the type rather than just the rate of change disjunctive rather than smooth? Will the success of a new product or approach make it necessary to replace rather than upgrade the existing one?

Quality of Suppliers.   There are many areas where the basic concept is proven, but few, if any suppliers can provide reliable facilities and equipment. The supplier may be a vendor of equipment or a PTT. The ability of PTTs in particular countries to provide services varies widely and may continue to do so.

Delivery and Service.   The willingness of vendors and PTTs to provide service may be a constraint. They will be stretched to their limits in many cases. They may not be able to meet delivery commitments, avoid backlogs, and provide support for installation and maintenance.

There is no easy way to decide where, when, and for how long the availability of particular components of the overall telecommunications technology base will be a constraint. To a large extent, the choice to view some component as a constraint and not as a design option reflects the policy on risk and on planning horizon.

Regulation

Regulation is a very real constraint on telecommunications.

Technology.   In many countries, the PTT has control over the technology itself. Many U.S. managers underestimate this and assume that telecommunications must follow the same trend as in the United States, toward increasing liberalization and competition. There is plenty of evidence to indicate that this will not happen in many countries. PTTs and governments too see telecommunications as a major opportunity and want to increase, not reduce, control over it.

Industry.   Deregulation has been a major incentive for firms to use telecommunications for market innovation. Regulation is a corresponding blockage to doing so. It is essential to decide if regulation is to be viewed as a constraint or a design option (or as an area for lobbying), and to anticipate who the firm's competition will be if it changes.

Geography. Regulation of international telecommunications compounds the constraints imposed in individual countries by industry and technological regulation. The issue of transborder data flows is especially complex and contentious. Sweden, for example, requires an export license for any data that may identify an individual (Scandinavia has among the world's strongest regulations on privacy).

As with the availability of technology, the stability and predictability of regulation is a major issue for planning. Here again, experts disagree: the legal issues are complex, with few precedents and a change in government economic or social policy can lead to important changes in telecommunications regulations.

International regulation is likely to be the most binding constraint. U.S. telecommunications managers in multinational firms generally do not get to know the situation in other countries nor get out into the field enough. The United States is totally unrepresentative of the international environment of telecommunications.

Industry Practices

A subtler constraint concerns intraindustry and interindustry cooperation. For example, the airline industry and international banking have well-established standards for key documents and procedures. Telecommunications as a service rests on interconnection, commonality, and standardization. Competition thrives on their absence in many cases. IBM for a long time explicitly prevented interconnection of other vendors' equipment with its own.

The customers in a given industry may force cooperation, as with IBM's gradual acceptance that customers required it to accommodate non-IBM facilities and most airline carriers' acceptance of tickets issued by other ones. Will there be common standards or shared facilities for trade documents, point of sale, payment mechanisms? To what extent must we accept existing practices?

Especially in a newly or semideregulated environment, telecommunications can quickly change industry practices. Customers become competitors. Competitors cooperate. Suppliers cut out the middleman. No one can reliably anticipate such shifts in industry practice, particularly when the forces for and against deregulation keep shifting direction. Telecommunications is stimulating a new form of vertical integration such as Dun and Bradstreet's purchase of the Thomas Cook travel agency firm in the United States to complement its on-line Official Airline Guide and reservation system. It is also leading to joint ventures between firms that previously had no reason to link up with each other: IBM and Merrill Lynch, and Citibank and McGraw-Hill are examples. These two failed, but what new ones will emerge and work? What policy or strategic alliances should we adopt now?

The Firm's Capability

One constraint that will undo the strategies of some firms is simply the shortage of good personnel. Very few organizations have anywhere near the quality and number they need for an aggressive expansion in their telecommunications resource. In this field, even more than in data processing, the bottleneck to innovation is not technology but people—for management, development, maintenance, and operations.

It has always been hard to recruit and retain capable managers in the data processing field, especially the hybrids who understand both business and technology and who are experts in new technical and application areas. The situation is worse as far as telecommunications staff are concerned because the pool of talent is far smaller, the rate of growth in the scale, scope, and complexity of investments in telecommunications more recent and more explosive, and the pipeline of people coming out of universities inadequate to meet demand.

Chapter 9, on organizing for telecommunications, discusses the types of people needed to manage the telecommunications business resource. In many instances, the only way companies will create the cadre of talent they need is simply to poach from other firms. Every airline is vulnerable to losing key personnel to banks; the airlines have the longest experience of any industry in running large-scale networks. The banks need that expertise, and when the network is the franchise, they are willing to pay whatever they have to. The Federal Reserve Bank system lost a number of its best staff when American Express geared up its communications activities in the early 1980s. When Bank of America decided to invest several billion dollars to catch up in electronic banking, it recruited Max Hopper, the star of American Airlines (he didn't stay, in fact). Only the best is good enough when telecommunications becomes central to the business vision.

If organizational capability is already a constraint on development, management, or operations, the firm cannot afford to push the limits of technology or extend the scope of its commitments to telecommunications. It has to decide if this is just a short-term problem that can be solved by hiring or education. It must also make sure that the constraint is not increased by any of their own key people being lured away.

DESIGN VARIABLES

Today's versus Tomorrow's Needs. Would we rather err on the side of having extra capacity or capabilities we do not in fact need rather than err by keeping costs down even if it turns out that that means not being able to meet demand for services? The answer for one engineering company might be "Our business vision rests on being the low-cost producer in our industry; cost is the priority." But for another it is "We view the network as the key to integrating design, manufacturing, and distribution. We must be able to expand services and volumes fairly quickly; we cannot afford to run out of capacity."

Security. Do we have a policy on security? If so, the question is answered. If not, what trade-off will we make between security and cost? Is security sufficiently important from a business viewpoint that we would not use or offer a new facility (such as a public data network, a satellite service, videotex, or customer service databases) if it lacks adequate security?

Performance versus Reliability. We can choose to put our effort and investment into building a capability that provides innovative services. It may be unreliable for a variety of reasons: Overload at peak times leads to poor response times; the technology is complex and new; there is an inherent level of errors in the technology itself. ("Bit error rate" is the rate at which a single zero or one may be inaccurately sent or received. One error for every 10 million bits sent is typical.) How important is reliability for us? For electronic mail, an occasional delay or an incorrect transmission is acceptable, but certainly not for electronic funds transfers.

In this last instance, the trade-off between reliability and performance or reliability and cost, or reliability and speed of response requires a clear business model. It is impossible to answer the question, "How reliable should our network service be?" We have to ask "Given these application opportunities, these policies, and these external constraints, what is the relative emphasis we should put on reliability? How much would we pay to improve it by X?"

The technical literature on telecommunications design emphasizes network "optimization." Generally, this means meeting given requirements (locations, number of workstations, volume of messages) at lowest cost. Minimizing cost is sometimes the key objective, and it makes no sense to pay more for something rather than less, especially since telecommunications is expensive in terms of both investment and operations; however, cost is not the only priority. Telecommunications for business strategy involves multiple, often conflicting objectives. We would like to maximize service, control costs, guarantee ease of expansion, ensure complete security, and so on.

This means there can be no "optimal" network design defined by purely technical considerations. Figure 7–3 shows the categories of technical and financial variables that have to be balanced with each other to arrive at the best practical design. *Capability* defines the range of services to be provided, and *flexibility* the extent to which technical components can be changed or added to accommodate needs for new services, increases in volumes, and so forth. *Quality of service* includes reliability and response time. It is not the same as capability, which indicates what the network can do; quality of service determines the extent to which that capability can be depended on.

The difference can be easily illustrated. A telephone hotel reservation service allows customers to make bookings—that is the capability. If the phones are not adequately manned, customers will not get through or will be left on hold. The hotel is well advised to improve the quality of service rather than the capability in this situation.

The trade-offs between aspects of capability, flexibility, and quality of service can be complex. Fairly obviously, it is more expensive to provide a high-capability, extremely flexible, and very responsive facility than one that delivers a limited range of functions at a reasonable level of service.

That is why many companies have 10, 20, or even 50 different network facilities, each dedicated to a particular application, and all of them largely incompatible. Many of those firms are trying now to integrate those networks. When the customer workstation becomes the delivery point for a range of services, and more and more mail,

Figure 7–3. Telecommunications Design Variables.

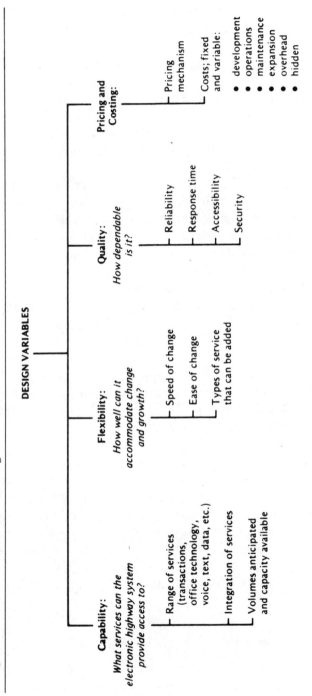

meetings, reports, and conversations depend on digital communications, high capability, flexibility, and quality of service become key. But what is the premium the firm is willing to pay?

The trade-offs involved in designing a communications capability are very complex and the technical issues more complex by an exponential degree. Figure 7–3 is not the base for technical design, but for managers to get a sense of the broad factors the designer has to address and, less broadly, to understand the relationship between costs and design trade-offs. Adding capability, flexibility, and reliability generally increases overhead and investment.

What *must* we have in these areas? What can we ease back on? What premium or saving in investment and operating costs are involved? These are obviously questions that require business directives. They illustrate why telecommunications planning must be done top-down, not bottom-up. They are impossible to answer if the business criteria are not already well defined.

The cost range for comparable networks in comparable firms can vary by a large factor. Some of the differences relate to management skills, economies of scale from being able to use fixed-cost leased facilities instead of incremental volume-sensitive ones, or technical and operational efficiency. A first-rate technical team can reduce operating costs to 40 percent below those in more typical and less well-run telecommunications units; this figure applies to the leading international banks and seems a general measure of the economies of expertise in telecommunications management.

Many of the cost differences, however, relate to design trade-offs. For instance, if the firm wants to be able to interconnect to customer networks and installs protocol converters that can convert messages from the format used in IBM's System Network Architecture to that used by "packet switched" public data networks, there is around 20 percent additional message overhead. This can affect response times, capacity needs, and so on. Is it worth it? Certainly it is for any company whose application opportunities partly or largely depend on being able to connect to customers' or suppliers' processing base.

Capability

Capability refers to the services and volumes that can be handled:

Range of Services.   These cover every type of computer-related traffic, from electronic mail, customer transactions, access to databases, word processing, videotex, electronic data interchange, voice mail, funds transfers, facsimile, etc. They must be based on the application opportunities.

With bottom-up design, they are quite often based on technical opportunities. Because videotex is available and experts predict it is on the verge of impact on business and consumer markets, or because local area networks are the fad-of-the-month in the communications and computer trade press, companies will often reverse the sequence of means and ends. They implicitly ask "Here is a solution. What is a problem for it to address . . . ?" rather than, "Here are our aims and priorities. What are the right vehicles to use?"

Of course, they have to look ahead and ask in addition which new vehicles such as videotex or local area networks might open up new application opportunities. The business vision is the anchor and reality test for answering that question.

Defining the types of service to be accommodated in the design of the telecommunications resource is in no way the same as deciding that they must be incorporated now. This is the difference between the architecture, which is the framework for evolving new facilities and services, and the topology and configuration, which is the operational system at any point in time.

Integration of Services.   The types of service to be accommodated in the architecture need not be integrated. They may require separate terminals, use different transmission facilities, and even when that is not so, be entirely independent of each other. For instance, a word processor may also send and receive electronic mail but not be able to process transactions. A customer workstation may need different procedures for getting inventory data, placing orders, and checking credit information. A personal computer may do all of those functions but not include a digital voice facility—it is a terminal and not a "telset" (which combines terminal and telephone).

The concept of "integration" in telecommunications and computing is elusive and at times rather like the advertising slogan "New and improved." From the perspective of the senior manager, the issue is not integrated technology but integrated customer service. What range of services can the customer get access to from a single workstation? Can separate transactions be cross-related and handled together? For instance, a travel agent may be able to make plane and hotel reservations from the same terminal but have to make separate calls with no resulting master travel record. To change or check on the reservation requires two more transactions.

From the technical specialist's perspective, integration relates to more shared facilities. For example, voice/data integration means that the same transmission links and communications switches handle telephone and computer traffic.

The senior manager's interest is in the services that can be delivered to the same workstation and combined together to create a new business opportunity. There may be many technical reasons for a firm to move ahead on voice/data integration but no business ones.

One key aspect of integrated services is interconnection to other firms' or industries' communications networks. EFT/POS is an example of this. It links the retailer's and the bank's network. They remain separate, but a data switch creates an integrated customer service.

It is far cheaper and easier to design a network capability that provides specific services than one that plans ahead for full integration. It may also be that customers do not need integrated financial services, integrated software that combines spreadsheets, word processing and database management, or integrated terminals.

Or they may. When? Why? The issue of the degree on integration that the architecture should provide for is again a business question that cannot be answered by forecasts, only by vision and the act of faith to make radical business moves.

Volumes and Capacity.   Telecommunications capacity is expensive. It is also not defined in a single measure but involves interdependent pieces of equipment, each of which can become a bottleneck that affects the whole system. For example, a reservation system may have ample transmission capacity but the switches—the equivalent of the telephone exchange—cannot handle enough terminals at the same time. The switches and the transmission may be more than adequate but the computer system they access is too small. A communications-based service is only as strong as its weakest link.

Forecasting capacity needs is close to impossible. With telecommunications, supply creates demand, and the past is no guide to the future. If a service is new, convenient, and useful to customers, and profitable to the supplier, then the firm surely hopes they will use it a lot, not a little.

Companies consistently underestimate the rate of growth in telecommunications traffic, and their short planning horizons and concerns to control communications costs often put them in a reactive position. One major bank lost market share in international cash management because of this; the demand was there, the capacity was not—which meant the service was not.

Should the firm err on the side of overestimating its capacity needs and risk carrying excess costs or underestimating them and risk not being able to meet demand? The question does not have an absolute answer. It can be avoided to some degree by making sure the architecture has enough flexibility to add capacity quickly, but that, too, may involve paying a premium. It may also be handled through the pricing mechanism, to encourage or inhibit growth and in effect ration capacity. This makes pricing and costing integral aspects of design, to be traded off against capability, flexibility, and quality of service.

Flexibility

Can changes be made easily to any or part of the electronic delivery base, including the computers and processing software? What range of changes can be made without having to replace, redevelop, or throw out existing components?

As mentioned before, one way of reducing the problems and risks involved in anticipating capacity needs is to choose an architecture that allows fast and easy change and that makes it possible to add capacity in small increments. This is a strong argument in favor of using public data networks or carriers such as MCI and GEISCO, which generally supply large amounts of transmission and the expertise to plan, install, and operate it quickly.

Here again, though, there is a trade-off to make. A private network may be more secure than a public one, can offer economies of scale, and can be optimized to handle the firm's traffic mix. And MCI or GEISCO may not be able to help much when the bottleneck relates to computer hardware and software.

The most flexible and general architectures are, by and large, the ones that have most overhead and hence add cost. That is the argument for and against adopting IBM's key architectures and computer operating systems. What do the senior managers want their telecommunications staff to guarantee: the minimum practical cost or the maximum practical flexibility?

Quality of Service

Regardless of what the capabilities of the telecommunications resources are, how important is reliability, response time, and accessibility?

Reliability.   How often will the customer get a busy signal, especially at peak hours, or find the system is "down," or encounter the problems that are just irritants on the telephone system but can be disasters when they create errors, delays, and disruptions? Noise on the line may lead to "$26,000" being sent as "$2,000"; lines, messages, or records may be lost; or the customer may be told by the computer "The system will not be available today due to maintenance." Reliability is essential when the network is the business. It is also expensive to ensure.

Response Time.   How quickly will the customer's message be processed and an answer provided? As with "integration," "response time" has different implications for the technical planner than for the business planner. The former has to move traffic along the highways through a series of lines, switches, front-end processors, concentrators, protocol converters, computer operating systems, and the like, under varying traffic conditions, including the 2:00 P.M. rush hour when every corporate customer wants to move money into and out of its bank. Tiny delays and additions to overhead, measured in milliseconds and bytes, can mean severe degradation in service. Planners aim at service levels such as "95 percent of all transactions being handled in under 3 seconds."

Sometimes, from a business perspective, it may be more important to guarantee only 6-second response for 99 percent of the time. Customers often object to the unpredictability of the response time rather than its specific average length. Sometimes, they may not need response in seconds. One major bank spent millions of dollars to give international customers a 5-second response for funds transfers then learned they were perfectly happy with 20 minutes.

Accessibility.   How simple and convenient are the procedures for customers to get access to the service? Do they need special equipment? Can they use the facility only within given times and from given locations? Must they use complex procedures?

There are many corporate employees who do not use their firms' new telephone with callback, autodial and the like, nor the service that cuts phone bills by 20 percent, because they have to dial too many digits first. Many customer services are attractive but inconvenient and cumbersome. The value of telecommunications-based delivery depends on the usefulness of the service, such as check cashing on Sunday, plus the usability of the system. In general, most people weigh usability more heavily than usefulness.

This is an area where communications and computing come together. Many aspects of usability relate to the design of the terminal, the procedures for logging on to the system, and the quality of the network facilities. These fall under the responsibility of the telecommunications unit.

Many other aspects depend on the design of the system software and on human engineering. These are generally handled by the information systems function. The term "user-friendly" has become a catch-phrase for requirement in designing systems for nontechnical and casual users; what it really means is "not as user-hostile as before." Computers are still hard to use. Procedures are like an intelligence test where the penalty for mistakes is to see your card fall into the maw of the machine. Accessing the network feels like playing Russian roulette.

Electronic delivery of services relies on accessibility, convenience, ease of use, and freedom from worry. The telecommunications specialist may not have the time, interest, or understanding of the customer to ensure these are provided. Someone has to.

Security is the fourth main aspect of quality and service. For telecommunications, security has to cover the whole chain of activity involved in electronic delivery. For instance, the communications lines may be "encrypted"; this means that the message is sent in a scrambled form that makes it look like garbage. Encryption eliminates many risks of breach of privacy, fraud and accident, but if the user's ATM card and secret password are stolen and used, encryption is of no help.

Companies and customers have, on the whole, been remarkably casual about security, though the credit card industry has become less so recently, because of the levels and costs of fraud. Simple passwords, encryption, and cards with magnetic stripes cannot provide adequate security. Visa found that about 30 percent of customers forget their password. That is why we all tend to use ones we can remember—spouses' or children's names, for instance—or write them inside the back of our diaries. Convenience and ease of access are always likely to get in the way of security and vice versa.

Should the firm's architecture be based on a policy about the type and level of security? Its managers must ask. Should it accommodate likely future needs for any tools to ensure security, such as "smart" cards and electronic signatures? How much security must we have? When is security a differentiator or even a product in itself? How much are we at risk from errors, omissions, and fraud?

Ensuring good security is very expensive. Today's electronic services are not at all secure.

Costing and Pricing

The Achilles' heel of many effective telecommunications strategies is the pricing mechanism. Choosing it is as much a design variable as deciding on the volumes to be accommodated. In fact, the two issues are interdependent. A major European computer manufacturer, for example, set up a dealer network service that accommodates about 2,000 terminals and 30,000 transactions a day. There are currently only 200 dealers using the pilot service. The firm has many options in pricing the service, each of which is likely to lead to very different patterns of use:

- *Give it away.* The marketing staff argue that if every dealer uses the system, the firm will benefit in terms of business growth, image for responsive system, dealer loyalty, etc.

- *Charge a subscription fee.* This is like paying an annual road tax for one's car. The dealer is charged $100 a month. (But too low or too high a fee can badly affect the business value of the service and the firm's costs.)

- *Charge by usage.*

The pricing mechanism for communications is a key determinant of demand. Too often, firms base prices directly on costs, by budget allocations for instance. Costs can be hard to define; how is a multi-year investment to be amortized and included in the cost base? Where are the hidden costs, such as support for remote users? Are computer software and operating costs to be handled separately from direct communications costs? The central issue for both pricing and costing is surely How do we want the user of particular systems to behave?

Pricing and costing mechanisms are design variables, or at least should be. In firms where there is a preset allocation method, pricing and costing have become a policy choice that may preclude some design options. That is why this is the Achilles' heel of many excellent technical strategies.

A Cautionary Tale.   In one of the top U.S. banks, use of an electronic mail system went from 400 to 4,000 in four months. That sounds like a success story. Most of the growth, though, was from messages sent from New York to the Far East to place football bets legally. The system was free and people responded not irrationally: they used it as they wanted. In the same bank, usage of a potentially valuable marketing database declined. The costs were based on fully allocating in the first year the capital cost of the software. This made the unit cost high. Usage dropped, which made the unit cost even higher.

OPERATIONS AND MANAGEMENT VARIABLES

The final category of planning variables shown in Figure 7–1, network operations and management, is of more concern to the telecommunications manager than to the senior executive, except when it has been neglected with the result that the system performs inadequately.

A large firm's private telecommunications utility is extremely complex to run. It is the direct equivalent of managing one's own electrical utility and telephone company. There are many advantages to having someone else run it instead, by using public networks rather than private ones. There are corresponding disadvantages in this, however, such as security or loss of control. With a public network, the company cannot be sure that the growth in traffic from other users will not degrade its own service, and the company will still have to acquire and operate a lot of equipment on its own premises.

A vast range of operations and management variables has to be provided for to avoid business failure from failure in operations:

Telecommunications Operations and Control Center (TOCC).   Every network facility needs a control and command center that oversees the operation of the entire system. This is the real decision making unit at peak times when the business is on line.

Monitoring and Testing Software and Equipment.   Real-time operations make it essential to be able to monitor what is happening anywhere across the entire system, diagnosing faults and troubleshooting.

Backup. If any major part of the system fails, there must be provision for backup and restarting.

Field Support. When users have problems or, say, a multiplexer of PBX in one of the remote locations fails, the utility has to respond quickly.

Few firms have a coherent network management strategy. They generally rely on vendor support and service and manual repairs. This is unacceptable in a multi-vendor, multi-technology environment where the firm's entire cash flow is on line. One hotel chain calculated the real economic cost of its network being down at peak times at $50,000 a minute. Leading banks and airlines regard 99 percent reliability and availability as barely acceptable. The best competitors in the electronic marketplace are moving rapidly towards automated network management, with IBM's NETVIEW and NETVIEW/PC products emerging as the key elements in their strategy.

Network management is currently an art form. It is also expensive. Does the firm really want to be in the utility business? If it does or simply feels it has to be because it needs a large-scale or private telecommunications network—there is no point in designing a first-rate architecture and providing valuable services if the operational efficiency is not close to perfect. Perhaps the best advice one can give to companies is:

- If you can't run it, don't build it.

- If you build it, put in the resources needed to run it.

- A utility has to provide 24-hour guaranteed service and immediate field assistance.

Sometimes, communications specialists get locked into a very narrow mindset, become unable to see beyond the world of operations and do not relate well to business needs. That is understandable. The policy for communications may be clear, the applications that use the facilities well-designed, the architecture flexible and effective, but when network operations fail, all these become irrelevant.

FROM STRATEGY TO CABLES AND BOXES

Figure 7–1 can be read from left to right along a spectrum from business to technical priorities: Application opportunities obviously are business-driven and the design of the communications facility depends on them. Policy choices are organizational in focus. Constraints mix environmental business, political, and technical issues and internal organizational ones. Strategic design variables are mainly technical, with some organizational factors involved, but depend strongly on business needs and priorities.

It makes little sense to plan from right to left, to allow existing operational requirements to constrain design, to let design implicitly determine policy and that, in turn, to limit application opportunities. Telecommunications planning has to move from strategy to cables, not the other way around.

That said, cables can open up new opportunities for strategy. A well-run utility can provide the chance to substitute electronic customer delivery and internal coordination for mail and physical transportation. A flexible, high-capacity highway system can allow management to look for new services and products that create a competitive edge. A new policy on funding and longer planning horizon opens up new application opportunities.

The framework presented in the figure is not a planning methodology. Telecommunications is far too complex to fit into any simple scheme. It is a framework, though, for relating the many variables that comprise a real strategy for network planning: competitive, economic, technical, and organizational. It is the basis for a dialogue between first-rate business people and first-rate technical specialists. Telecommunications planning involves many tradeoffs and it is essential to have a clear business model for making them. Every one of the wide range of issues requires in-depth analysis and evaluation for any given company. As one moves from left to right in Figure 7–1, the amount of technical and financial detail involved grows exponentially. The simple message to managers is:

- Focus on opportunities first.
- Be explicit about policy.
- Identify the constraints on design and operations.
- Highlight your trade-offs for design.

The technical people can then do their work. The message to telecommunications specialists is:

- Know what the firm wants to "optimize."
- Highlight your need for policy directives.
- Don't get locked into a technical, operations-centered mindset.
- Talk to the business people.
- Highlight the design trade-offs.

The dialogue between business managers and technical specialists should be about trade-offs.

The missing dimension in telecommunications planning has often been a lack of policy. The missing resource, with or without policy, has been authority. Policy must be backed by authority and accountability: Who decides and who is to be responsible? Authority and accountability must be meshed. Integrating business thinking and technical planning needs a new style of leader for an integrated information function. Unfortunately, many telecommunications and data processing managers do not have the business interest and understanding or the skills needed to lead. Senior managers are in the role of the princess who with a kiss can turn the frog into a handsome prince. Sometimes, though, when they kiss the frog, it turns out to be just a frog. Then the princess needs to place a "prince wanted" job ad.

8 MAKING THE BUSINESS CASE FOR CHANGE

Traditionally, investments in data processing and telecommunications have been justified mainly in terms of cost displacement. This largely reflects a historical focus on automating clerical operations. It is also in part a reaction to the often uncontrolled growth in expenditures on data processing in the 1970s and the many unmet promises about the payoff. Senior managers are wary when invited by the information systems department to spend just a few million dollars more.

When telecommunications was almost entirely a telephone and telex utility, cost was the main issue. Service levels had to be maintained, but there were few "value-added" benefits to be gained from the available technology. As telecommunications became a force for innovations that affect business survival it became hard to justify the new investments in old ways. Obviously, though, those investments have to be systematically evaluated and justified. One cannot make the business case for major expenditures on telecommunications by vague talk about "productivity" or "competitive edge" nor underplay the very real risks involved in applying new technology to unproven business opportunities.

Vision and policy again are the starting points. There can be no one way to make the business case. Two extremes are to be avoided: forcing inventions and innovations to be put into a straitjacket of cost justifications and evading the very real problem of how to define business value when payoffs cannot be precisely forecast or benefits quantified.

An innovation like Merrill Lynch's CMA is a radical move. The risks are high. The volumes and benefits depend on customer acceptance. There are few precedents to measure it against. No amount of cost-benefit analysis is likely to convince a skeptical or even a neutral listener. The case has to be made in terms of business trends and market opportunities.

The radical move has to be justified to and by the people at the top of the firm, and the time horizon they must address is a long one. The operational move requires a shorter, more budget-centered view, with decision makers at a lower level.

Rather than think in terms of "the" business case, we need to recognize there is a range of business cases suited to the spectrum of telecommunications proposals, which goes from operational to incremental to innovative to radical moves.

- *Radical moves* explore an uncharted business, technical, or organizational area; involve major new commitments and risk; are an act of faith with no guaranteed payoff; and will have a strategic impact on the direction of the business if they are successful.

- *Innovative moves* involve a substantial increase in the level of commitment; build on current capabilities and experience; are based on a proven concept, though there are significant risks in implementing it; and provide substantial business value if successful.

- *Incremental moves* are the "next step" in an existing application within proven technology, involve an increase in scale or scope; have some degree of risk, and provide a measurable return on investment.

- *Operational moves* are a better way of providing an existing service or handling a standard operation; substitute new technology, equipment, or procedures for old; and involve limited, manageable risk; contribute directly to the "bottom line."

Making the business case involves getting the attention of the right people, marshaling convincing data, and persuading the people to make a choice. The people, the data, and the nature of the choice differ for each type of proposal. Figure 8–1 summarizes the sequence of steps.

In a radical move, what this book calls the "vision" is the business case. It sees communications as a means of refocusing key parts of the business. It generally implies substantial potential benefits.

Figure 8-1. Making the Business Case for Telecommunications.

Classify the Proposal How radical is the proposal?	Identify "Panel of Judges" Who can make the decision, and whose support is also needed?	Identify Data Needed to Make Case What competitive and/or financial information will be most convincing?	Provide Initial Briefing/Education What does the panel of judges need to understand?	Present the Options	Analysis of Each Option
Radical Move: Communication as a means of refocusing major aspects of the business	Senior business management and corporate business planners	Competitive positioning and long-term business growth	Raise consciousness; focus on competitive context	• Do nothing	• Business benefits
Innovative: Support for key business activities in new ways	Business managers responsible and financial staff	Impact on revenues and profits	Highlight business options and implications for business planning	• A few intermediate alternatives varying cost and risk	• Technology and technical risks
Incremental: Logical next step in extending or improving an area of business	"User": management and financial staff	Return on investment	Provide briefing on content of proposals; highlight tradeoffs	• "Ideal"	• Financial
Operational: Improvement in cost, efficiency, or control	Financial and technical staff	Cost impacts	Provide detailed technical and financial vocabulary needed to evaluate proposal		

Decide

An innovative move might be a pharmaceutical distributor's building the infrastructure for managing inventories and putting terminals in the customer's store, in the same way as did McKesson and American Hospital Supplies. The concrete business picture may well come from looking at what other firms are doing.

Incremental moves fit within the existing vision. For the grocery chain Publix, the decision to compete with Florida banks for control of EFT/POS was radical; the decision now to add a new bill-paying facility to its ATMs would probably be seen by its managers as incremental. Here communications adds a service or does something in a better way.

By definition, an operational move also fits within existing planning assumptions and organizational procedures. It does the same thing better. In the case of Publix, this might be an upgrade of the ATM equipment. Obviously, there is no point in even trying to justify a radical move if the vision is not clear and has not been successfully communicated by management.

CHANGING THE RULES OF THE GAME

Policy sets the organizational rules of the game. It defines authority, procedures, and planning assumptions—the constraints within which the business case must be made. It establishes territorial and political boundaries and, above all, delimits spheres of discretion.

The more innovative the move, the less likely it is to fit existing policy. An example is the proposal in a shipping firm to build an integrated communications capability that will handle all feasible aspects of document transfer, electronically link into banks' cash management systems to speed up the complex process of credit payment, and provide certification of ownership of goods. This intersection of electronic document management and cash management involves very complex legal issues and fairly complex technical ones. It costs about $500 to process the documents involved in a typical international trade transaction and a single week's extra delay in getting documents to the right place adds as much as $10,000 just in interest charges on a shipment of 200 cars or a tanker of oil, so that the potential payoff is huge. The business logic of such a proposal for electronic trade management is clear; however, such a radical idea—radical for all but the few firms in the industry that have already shaped such a vision—

does not fit in with established lines of authority and responsibility or with capital investment and budgeting procedures.

The telecommunications manager is not expected to propose major changes in business and marketing. He or she is expected to submit investment proposals through the regular planning cycle, and expenditures will be allocated to operating units through a cost recovery formula.

In practice, in many firms the communications manager would not even try to present such a proposal. It makes no sense to stir up trouble by going around established procedures and by intruding on other people's territory. A firm's policy for telecommunications should make it acceptable for technical units to present proposals that have a strategic business impact.

Level and Mode of Justification, and the Panel of Judges

The business case is presented to a panel of judges—real people who have to be persuaded. Some of them are not part of the formal decision-making process but their opinions have informal weight. Before designing the presentation of the business case, one has to ask

- Who can *make* the decision; which group or individual must provide the authority, resources, and commitment needed to turn this idea into action?

- Who are the *opinion leaders* these decision makers rely on, to whom they turn for formal or informal review? These people can provide support and influence for the move.

- What *approvers* must one have? These are the people who can in effect veto the proposal, who must at least indicate their nonopposition if the proposal is to have a chance of being accepted.

- What kind of *data* is needed to make the case?

Generally, the decision makers will be a business manager or committee of managers, and the opinion leaders, respected planners or outside advisers, and the approvers financial and technical staff. The exact composition of the panel of judges depends on the type of move and the specific organization. As Figure 8–1 suggests, as one shifts down the spectrum from radical to operational moves, the relevant data needed to make the case become more specific and cost-focused. At the extreme of the scale of innovation, the key data presented to senior decision makers are trends and pictures from the industry and marketplace. At the other end of the spectrum, the relevant data are precise analyses of cost and volumes, to be reviewed by the people whose budgets and operations are most directly affected. This point is central. Appropriate data must be presented to appropriate people.

Briefing and Education.   Telecommunications and business strategy is a new field for everyone. Managers lack the experience and the training to handle it and most of them never expected to have to make decisions about applying communications. If they have not been briefed well before the proposal is presented and do not feel comfortable in taking part in the evaluation process, it is too late to change the situation. Education has to start early and be sustained.

The needs vary. For radical moves, the aim is to raise consciousness and establish a business focus; the workshop and reports should be filled with examples from industry and analysis of trends in the industry; the technology is related to these, not the reverse. Innovative business moves need a sharper focus and a framework for planning; the business relevance of major options and trends in the technology has to be highlighted; the goal here is to help business people feel they understand the key planning issues. Incremental moves are focused on the proposal, not the broader marketplace and technology; one needs to highlight the relevant choices and trade-offs. For operational moves the vocabulary and concepts needed to assess the proposal must be clearly defined; the focus is on briefing not education, on specifics, not concepts.

The steps described so far (see Figure 8–1) all precede the actual presentation and evaluation of the proposal. They are all too often neglected but are essential in establishing the context, criteria, awareness, knowledge, and vocabulary. Without all this being done, how can complex new applications of risky technology and business payoffs that go beyond cost displacement ever be adequately assessed?

By far the most common questions communications managers ask in discussions of strategic planning are How do I get my senior managers to listen? and What do you do when they are only interested in the cost? The answer is to focus on vision and policy first, and build your education strategy well before you present your proposals. The questions can be restated to apply to senior managers: How should I be thinking about telecommunications? and What should I look at besides cost?

Presenting the Proposals

The uncertainties intrinsic in innovation make it essential not to get locked into presenting, and then having to defend, a single proposal. Equally, one must face up to and even highlight risk rather than downplay it. The more innovative the business application, the fewer the precedents against which to compare it. The greater the likely payoff, the greater the likely risk. The decision to make an innovative or radical move is a business decision about return and risk. The real issue is how far along the spectrum of risk and return management is willing to go, not whether to say yes to one specific combination of the two.

This is something too many data processing managers overlooked in the 1970s. They generally presented their senior management with a single proposal and rarely stressed the risks involved. This damaged their credibility and put them on the defensive. They had to respond to questions about risk with "yes, but" or discount the problem. To do otherwise meant the proposal being turned down.

A far better approach is to say "This is our preferred option. It has these risks and potential payoffs. Here is an alternative that has either less risk or less return. You choose." "You" means the panel of judges; the telecommunications manager has to give them a real business decision to make, not a go/no go response to a single proposal.

There is generally, for any one business opportunity, a wide range of technical options and levels of commitment. Obviously, it makes no sense to look at all the combinations. A much more effective approach is to narrow the analysis to three or four key alternatives:

- *Do nothing*, a real option, has to be evaluated systematically.

- *The "ideal"* alternative assumes the best of all worlds—things will work out as planned, market assumptions will be correct, the technology will work, etc.

- *A few options between these extremes*. One of them will almost certainly be the one recommended by the communications manager and should be flagged as such.

Based on these options the implications of changes in key factors can be analyzed from the start.

VALUE ANALYSIS

The formal proposal for any major investment must look at both cost and value. Generally, they are compared directly; however, until the value is established, *any* cost is disproportionate. Too often, as well, the costs are quantitative, predictable, and immediate and the benefits qualitative, uncertain, and deferred. Business inventions cannot be cost-justified, nor can direct and simple comparisons be made between cost and benefit.

Innovation is value-driven. There is a wealth of empirical data to support this point. The principles of value analysis are:

- Separate benefit from cost.
- Establish the benefit first (Ask: What is this worth?).
- View cost as a threshold (Ask: Would I pay $X to get this benefit?).
- Face up to qualitative benefits.
- Rank order "hard" and "soft" benefits.
- Define key indicators by which qualitative benefits can be evaluated.
- Use "bounding" to get a rough estimate of needed and likely benefits.

In most of the examples shown in Part II of companies using telecommunications to gain a clear business advantage, cost displacement was not an expected or intended priority. Because automation in general, and telecommunications in particular, have traditionally been viewed in terms of cost, it can be hard to shift the perspective. Organizations are far more skilled in cost analysis than benefit analysis. The way to begin to change this is to learn to list, track (through pilots), and rank benefits.

THE DYNAMICS OF INNOVATION

Innovations are made by organizations on the basis of value. The early adopters differ from the late ones. They focus on a small number of benefits, which usually relate to a specific "felt need" with a definite time horizon. They are rarely interested in "office automation" or "productivity" but in getting the budgets in on time, seeing where they can improve short-term forecasting, or keeping the sales force better informed about inventories and prices. Innovators are not risk-takers. In fact, they tend to be skeptical about wild claims and do not respond to "technology push."

Many surveys show that most product innovations come from "demand pull"—the definition of a need in the marketplace to which the technology is then adapted. Technology does not create needs, though education about its capabilities and availability may. Quite often, knowledge about the technology is brought to the innovator by "gatekeepers," who bridge the two cultures of business and technical expertise; academics and planning staff often play this role.

Technical specialists too often think in terms of technology push. That is a solution looking for a problem. The approach to telecommunications and business strategy recommended throughout this book is to focus on the benefits in terms that are convincing to innovators—businesspeople who combine healthy skepticism about technology and open-ended promises with the willingness to put authority and resources behind a project once they are convinced. These are the people who can change the terms of reference for telecommunications. There are three ways to do so:

- *Focus on exemplars*, as in Part II of this book; these are proven and concrete illustrations of benefits and opportunities.

- *Define the business vision* for telecommunications.

- *Focus on the benefit side* of the cost-benefit calculus.

This does not mean evading the issue of cost, but knowing where to put the emphasis. For radical moves, a strong focus on value is called for, with costs defined in terms of a floor: at least X, probably Y, and no more than Z; once the benefits have been defined and management is convinced, then a follow-on study can look at the costs in more detail. For innovative moves, emphasize value but also be more detailed in indicating cost components and ranges. For incremental moves give equal weight to benefit and cost, but treat them separately and in that order. And for operational moves, be very precise about costs, since that is the issue here; the firm does not need to make the change and the real benefit is the cost impact.

TECHNOLOGY ANALYSIS

The risk inherently involved in information technology should be highlighted rather than smoothed over. As with investment in stocks and securities, high return generally implies high risk. With telecommunications, getting a strategic edge means doing something earlier than one's competitors, which may involve using a technology that is not fully proven. Value analysis establishes the return from the investment. Business managers can assess the business risk, the likelihood of getting the return, but cannot intuitively do so for the components of technical risk, which include the technology itself, the product, and the vendor. Detailed risk factors which apply to all three of these are

- *Performance.* How likely is this to work as promised?
- *Delivery.* Will announcement dates and delivery schedules be met?
- *Substitutability.* If there are problems, can an alternative technology, vendor, or product be substituted without too much delay and cost?
- *Regulation* (especially key internationally). What is the likelihood of government (for example, PTT) regulation constraining acquisition, use, or even performance?
- *Control.* Are we dependent on the vendor or regulators if there are problems or changes?
- *Costs.* How reliable are our estimates of costs and cost trends? Are there indirect costs we cannot predict such as tariffs?

This list of risk factors is not complete, and the technical staff will have to go into far more detail in their planning and analysis. The goal here is not to produce an exhaustive technical analysis but to give the panel of judges a complete enough picture of the likely risks to be able to trade them off against the likely returns and to feel comfortable that they have been properly briefed.

FINANCIAL ANALYSIS

It is only after the issue of value has been established that cost can be brought in, and even for operational moves value and cost should be kept separate. This flies in the face of traditional approaches to cost-benefit analysis but is surely the only sensible way to handle innovative business proposals that rest on technology. Bundling costs and benefits together, trying to express uncertain qualitative benefits directly in terms of cost displacement or avoiding the problem entirely, simply muddles the issue. This in no way means one need not be rigorous about cost. The overall issue is handling uncertainty and establishing business value. This means be rigorous about value, be rigorous about risk, then be rigorous about cost.

Being rigorous about cost means focusing on cost dynamics rather than cost estimates. In most instances where a proposal for telecommunications introduces a new service, it is close to impossible to estimate volumes. In many instances, supply creates demand and companies consistently underestimate volumes sometimes because they previously overestimated them. An example is electronic mail.

Company A estimated demand would be fairly low; this was based on the assumption that 10–20 percent of phone calls and memos would be shifted to the new medium. It turned out to be very high in one division because the sales force found it an entirely new and effective way of keeping in touch with the office when they were on the road. In Company B, use was far lower than predicted on the basis of a pilot because project users were charged for the service whereas in the pilot study it had been free. In Company C, use was initially low but took off rapidly when a new software package was installed which made it simpler and easier to use.

Cellular radio, voice mail, full-motion videoconferencing, home banking, network information services . . . There is a flood of new technologies and new services where supply may or may not generate demand, cost and ease of use may change customer response, or peoples' learning may open up new applications. In many instances the hope is that demand will be far higher than predicted—that the productivity gains from videoconferencing will lead to the rapid substitution of telecommunications for travel, or that distributors will enthusiastically adopt the dealer order entry system, or that the new all-in-one reservation system will be a winner.

For all these reasons, it will often be impossible to predict volumes and hence costs. The two key questions then are

- How do costs change with volume?

- Would we rather overestimate demand and have unused capacity or underestimate it and not be able to meet demand?

The second question is a key business issue. It can be rephrased more specifically: We expect that about 200 customers will sign up; if we get over 1,000, we will be unable to accept them without significant degradation in service. If we get only 700, we will be operating at a loss. Which direction of risk do we choose—oversupply or under-capacity?

Costs do not necessarily (or often) increase directly with volume. There are many components which have a fixed capacity, and expansion involves additional capital investment. For example, a communications switch can handle a given number of terminal "ports"; when that limit is reached, new capacity must be added in multiples of ports. The dynamics of cost are thus shown in Figure 8–2.

The costs may be significantly affected by changes in tariff structures and rates. The use of private facilities and satellites may suddenly become less cost-effective when a foreign PTT or U.S. supplier switches to "volume-sensitive" tariffs or reduces costs for terrestrial circuits. This is a growing problem; not only can we not predict volumes, but we may be wrong by a large margin in our assessment of cost trends.

The problem of projecting costs will not go away by ignoring them. We need to summarize them in terms of "If the capacity is X and the volume Y, the likely total cost is Z." The issue is not cost per se but net business value. For a firm that wants to use its communications capability to deliver products electronically, cost is only part of the calculation. But in each instance, the business questions are:

- What is the likelihood of particular traffic levels?
- What happens if the estimate is wrong? Would we rather err on the side of having too little or too much capacity?

Only management's panel of judges can choose the preferred risk direction. That is a central issue for telecommunications when the traffic has significant business value, or the relative costs are high, and in either instance volumes cannot be predicted. The telecommunications manager can highlight the technical risks and show the net business value for given traffic and capacity levels, but it is not his or her job to make the final decision, only to present the business case.

Figure 8–2.   Cost Dynamics: How Fixed and Variable Costs Move with Volume.

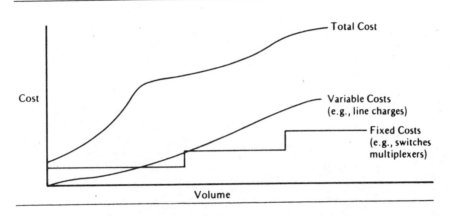

GIVE US A REAL BUSINESS DECISION TO MAKE

In practice, the business case is too often made in terms of cost and cost recovery, of single proposals, and of technical issues. That does not make sense, except for simple operational moves, and it locks out business innovation through telecommunications. Communications managers have to work through the stages outlined in Figure 8–1 and create a business-focused dialogue, as do business managers.

If there is no vision, how can existing policies be challenged and changed, so that opportunities for radical and innovative moves can be brought into top-level planning circles? If the policy for telecommunications cannot be changed, how can anyone step outside or around the existing justification process? If communications managers cannot influence the justification process, how can they find and convince the right panel of judges? If they cannot brief or educate the judges, how can they present their cases?

Senior managers can change policy and procedures far more easily than technicians. The strategy for making the business case presented in this chapter is the process that top managers can and should encourage, rather than the one that communications planners should maneuver to facilitate. The demands from senior executives to their communications staff then should be: "Highlight the uncertainties" and "Give us a real decision to make."

IV TAKING CHARGE OF CHANGE

Vision, policy, and architecture are the essential elements in planning for telecommunications for business strategy. In themselves, they are not enough, though, to mobilize the organization for the radical change telecommunications directly implies—change in organization, technology, management, and work. A constant problem over the past 30 years in introducing computers into both large and small firms has been how to make sure that new systems work both organizationally and technically. The problems are far more complex for an era of telecommunications as a coordinated business resource, because the scale, pace, and scope of change are accelerated.

The two focal points for change are the management process and the organization for delivering the new systems, and the strategic education process to help people adapt to and use them. These are the topics of Chapters 9 and 10 of this book. They are not about managing change, but about taking charge of change.

Generally, the organizational change stimulated by computers and communications has been viewed in terms of programs, which have a beginning, middle, and end. A project is defined around a specific application, such as "Build the on-line customer equipment system" or "Install word processing for a department" or "Implement a new financial reporting system."

The implementation process starts by building mechanisms for "user involvement." This requires real participation and mutual understanding between the bringers of change—the technical designers and project managers—and the receivers, the people who will use or be affected by the innovation. The need for involvement is one of the clichés of the systems development field. Making it effective, instead of pseudoparticipation, is less easy.

Involvement is generally complemented by training. That usually follows implementation and focuses on helping people use the new system and not feel threatened or lose a sense of competence and autonomy. The process of implementation may take several years for a large-scale system but at the end of that time it will be complete in two equally important ways. First, it has been thoroughly tested and is technically reliable and efficient. (There will *always* be some subsequent problems. There can be no entirely error-free system because of the astonishing complexity of detail involved. If a single computer instruction causes an error in an application, the entire system may fail. A large-scale central computer installation may typically encounter a "bug" in an operational system every three days. In that time, it will have carried out a trillion instructions. Perfection is not possible in large-scale application of increasingly integrated and, hence, interdependent technical components.) Second, the system is embedded in the organization and institutionalized. People have the skills they need; and problems of attitude, concern about jobs, career prospects, and quality of working life have been honestly addressed and resolved; and the system fits smoothly with reporting procedures, job incentives, mechanisms for planning and control, responsibilities, information flows, etc.

It took the data processing profession several decades to learn how to mesh the technical and organizational aspects of implementation, but there is now no excuse for the old types of organizational failure of technically sound systems. A complex project can be smoothly managed and information technology need not be synonymous with disruption.

Unfortunately, there will be fewer and fewer projects with tidy beginnings and endings. When a bank automated its back-office processing in the 1960s and 1970s, there was a stage when the main task was accomplished. The systems were complete and there was a period of several years when no major new changes were introduced. In the 1980s when that bank puts in customer service workstations and on-line databases, the process of adding applications, changing work, and redefining the organization will go on at least through the late 1990s. When a manufacturing company commits to computer-integrated manufacturing, there will be no single project, but waves of projects. As one ends, another is starting up.

Change is now the norm. That is perhaps the largest organizational impact that telecommunications is creating. This means that it will not be enough to create "involvement," provide training, and manage discrete projects well. "Managing" change is largely reactive. It implies that as we gear up to respond to business pressures and opportunities, we can develop mechanisms and provide resources for smoothing the transition. When the pace of change is constantly accelerating and it comes in waves, managing it essentially means a defensive process of adaptation.

How do we take charge of change and create an environment which is one of *changing* and learning how to deal with radical (not evolutionary) innovation in and through technology? There can be no glib answers, and only senior managers can stimulate the shift from reactive innovation and managing change to getting ahead of the change curve.

ORGANIZING FOR EFFECTIVE DELIVERY

One of the first issues these managers have to address is the nature of the organization that handles the delivery of the technical systems. This is the topic of Chapter 9. It shifts the emphasis from telecommunications to the integration of telecommunications and information systems. The integrated technologies now demand an integrated organization.

Telecommunications pushes computer terminals out into every part and level of a business, results in technical risk becoming business risk, and makes more and more work "computer-mediated." Managers, secretaries, the sales force, supervisors, planning staff, and engineers become direct "hands-on" users of workstations. Their jobs are in no way automated, but the workstation becomes the computer equivalent of the telephone—used perhaps only minutes a day but a central support to their work. Computers and communications no longer mainly represent radical—and risky—innovations in technology, but radical innovations in the nature of work.

The management process in most large organizations is not designed to handle such a situation. The history of data processing and telecommunications has been marked by delegation of authority, isolation of technical staff, and technocentrism, which explain most of the reasons for failures in implementation of information technology. Separation of technical elements such as transaction processing, data management, and office technology accounts for the problems firms are having and will continue to have in moving toward the integrated computing and communications resource they need to compete in the electronic marketplace.

Other historical problems in introducing new systems have been poor technical design, poor project management, and basic flaws in the technology itself. By and large, these last problems have been resolved. After 20 years of painful, often vicarious trial-and-error learning, we have effective techniques for design and the steps involved in the systems development life cycle are well understood.

More important, there have been astonishing and continuing improvements in and reduction in the cost of software packages, personal computers, data communications, and data management tools. In the 1960s the basic technology was hard to tame and make work. In the 1970s the bottleneck was managing large-scale software development projects. Throughout that period, the technology was the issue. Delegation of responsibility to technical specialists, isolating them from the wider organization, and defining management processes that centered around the technology itself were reasonable and natural responses to the problems of computers.

This is no longer a suitable reaction to the opportunity of telecommunications and the integration of communications, computing, and data management that it makes possible. From the truly awful history of data processing, we can predict the consequences of trying to manage the organizational change telecommunications is bringing by using the established management processes: disruption, late delivery, "resistance," and loss of credibility.

THE INTEGRATED ORGANIZATION

Integrated technologies require integrated thinking and an integrated organization. How to build that is among senior information systems and telecommunications managers' biggest challenges.

Chapter 9 outlines how a few (too few) organizations are making the shift. It is a new generation of information systems or business-oriented telecommunications leaders who are creating the momentum. The telecommunications unit then becomes just one part of the infrastructure for an overall information function. It is lack of senior management understanding of the human side of the systems organization that blocks those leaders. They need a new philosophy of organizing and in the end only senior management can sanction that.

The new philosophy is based on

- *Bringing together* the largely separate and fragmented worlds of data communications, voice communications, information systems, and data management

- *Clarifying the blurred issue of authority and accountability* so that the architecture has an architect

- *Recognizing the differences* between and equal importance of development and operations

- *Creating a new set of broad planning skills* especially around the architecture, making the business case, and financial management

MOBILIZING FOR RADICAL CHANGE

Managing the radical changes of telecommunications mainly requires policy, resources, and programs in three areas:

Education.   To lead change (unlike training, which follows it), education has to be pervasive and sustained so that people have the vocabulary, the understanding of the business message, and the insight into the planning process to be meaningfully "involved" and "committed." It has to have clear behavioral objectives. The issue is not what people at different levels of the organization—including top management—need to know, but what they must *do.*

Systematic Cross-fertilization of the Organization.   The firm needs to grow hybrids who can move between the worlds of technology and business. These new information services professionals, product designers and marketers, and business planners will also be the talent base from which the next generation of senior managers will emerge. For them, the ideas and experiences discussed in this book will be managerial common sense, and they may wonder why it was so hard for business people and technical specialists to think and work together in the old days.

Institutionalize the Process of Changing.   The process is crucial rather than the techniques of managing specific changes. The business vision is the guiding blueprint for this. It has to include an organizational vision, too.

The dynamics of innovation for telecommunications and integrated technology begin with vision. They end with sustained action. Sustained action maintains the momentum and makes management commitment and user involvement meaningful through behavior rather than memos and vague goodwill.

By definition, radical change is not easy to manage. When it involves radical innovations in business, in organizational processes, and in technology, all at the same time and at an accelerating pace, the risks are immense. So, too, are the potential payoffs from moving fast enough to get there before the pack but systematically enough to ensure that idea and implementation work together.

9 ORGANIZING FOR TELECOMMUNICATIONS

For the foreseeable future, the bottleneck in exploiting integrated information technology in general and telecommunications in particular will be the supply of good people, not the supply of the technology itself. This partly reflects the gap between the need and the supply which has always been a problem in the information systems field. Typical large firms have backlogs of development projects measured in mancenturies, rather than manyears.

Personal computers and "end-user" software packages have helped cut into the backlogs but created new ones of their own as supply creates demand. A rough rule of thumb is that 50–70 percent of information systems staff is working on maintaining or updating existing systems, 20–40 percent on enhancing them, and only 10 percent on developing new systems. When the tax laws change, or a business unit adds a new customer service, or IBM introduces an improved operating system, existing programs have to be modified. The effort can be huge. When the U.S. Postal Service proposed to change the zip code to nine digits, calculations of the cost to the Fortune 100 firms of changes to their basic processing systems were in the billions of dollars.

The problems of maintenance and backlogs will be with us for a long time. Software tools for developing large systems are improving, but productivity is increasing at a far slower rate than demands for new applications. This, though, is a relatively small constraint compared to the fairly sudden shift in the entire skill base required for the integrated technologies. If one knows a person's job title in information systems, one may have no idea what he actually does. Figure 9–1

Figure 9–1.   Changes in Skills Needed for Integrated Information Technology.

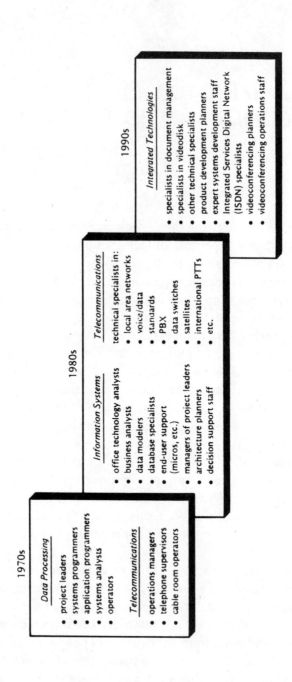

1970s

Data Processing

- project leaders
- systems programmers
- application programmers
- systems analysts
- operators

Telecommunications

- operations managers
- telephone supervisors
- cable room operators

1980s

Information Systems

- office technology analysts
- business analysts
- data modelers
- database specialists
- end-user support (micros, etc.)
- managers of project leaders
- architecture planners
- decision support staff

Telecommunications

technical specialists in:

- local area networks
- voice/data
- standards
- PBX
- data switches
- satellites
- international PTTs
- etc.

1990s

Integrated Technologies

- specialists in document management
- specialists in videodisk
- other technical specialists
- product development planners
- expert systems development staff
- Integrated Services Digital Network (ISDN) specialists
- videoconferencing planners
- videoconferencing operations staff

summarizes the changes from the era of separate applications to the portfolio of technologies and applications for the mid-1980s and the 1990s.

For the rest of this chapter, the abbreviations IIT and IS are used. IIT stands for "integrated information technology." IS stands for "information systems," the corporate development function for the main computing applications (the old term was "data processing" or DP, and "information management" is a more recent one). One of the problems in describing what is happening in the field is the lack of standard labels for and descriptions of the broadening range of jobs, skills, and roles it involves.

Figure 9–1 gives a flavor of the range. When we are short of people in the traditional jobs, the problem of building the new human resources base will be horrendous. It is worst in the area of telecommunications, especially in terms of finding individuals who combine virtually contradictory requirements.

- *Strong technical qualifications* in digital communications and in the integration of communications and computing: The ink is scarcely dry on the diploma.

- *Strong operational experience*: Older and with obsolescent knowledge, but solid understanding of large-scale commercial operations.

- *Proven management skills*: Something few technical specialists have the desire or even the instincts to acquire, wanting instead to build systems.

- *Knowledge of the business*: Where do technical staff get the lateral development and exposure to build the knowledge? Where do they find the time to keep up in the technical field as well?

Few people combine this mix of skills. They are at a premium in the marketplace and as firms recognize the importance of telecommunications to their business strategy, they are willing to pay whatever is needed to them.

It is not just managers who are scarce. Technical specialists in advanced telecommunications are always hard to find—and to keep—because the field is changing so quickly. Younger staff with Ph.D.s in computer science often lack real understanding of the operational side of telecommunications. The older ones need a sabbatical to update their knowledge base, but they cannot be spared.

The problem is not simply one of the limited supply of bright educated, motivated people, nor of salary. Over time, the market will alleviate if not eliminate that shortage. A far greater problem is that while a firm can go out into the market and bid up the price for first-rate technical talent, organizational experience has to be built not bought. Telecommunications for business strategy involves building the human equivalent of a wine collection. The intellectual crop of 1986 may be the best since 1888. But the wine has to mature for years.

Of course, not every technical professional needs to acquire those skills. But more and more aspects of exploiting the information technologies depend on the hybrids, people who combine strong technical and adequate business and organizational skills or strong business and adequate technical ones. These people have to be grown, not brought from the market. These are the human resources base for the new IIT organization.

They do not have career paths, because their jobs have never existed. They have only career trajectories—and immense career ambiguity. Figure 9–2 shows the problem.

IIT increasingly relies on taking people who are on a technical trajectory and moving them toward a business/organizational one. For example, the "information center" is an IS innovation, designed to help business units build their own development capabilities. They can use end-user software to build their own planning models and spreadsheet reports or design a small-scale management information system. ("End-user" is a vague term that basically means that any one with analytic skills can learn to use the tools; professional training and experience in information systems is not required.)

The skills required to staff and run an information center are very different from those of traditional IS development. Technical experience is far less important than personality. In many ways, the Infocenter is a marketing and consulting arm for IS. It is a service role. The best programmer may be ineffectual in the Infocenter; he or she is a professional with strong standards and focused skills having to support "amateurs" muddling through ad hoc projects, doing quick-and-dirties (QADs).

Managing the information center is as much an organizational as a technical role. What is the career "path"? The job has never existed before. The center's manager has been pulled away from the technical career trajectory and runs the risk of becoming a mediocre technician

Figure 9–2. The Problem of Career Ambiguity.

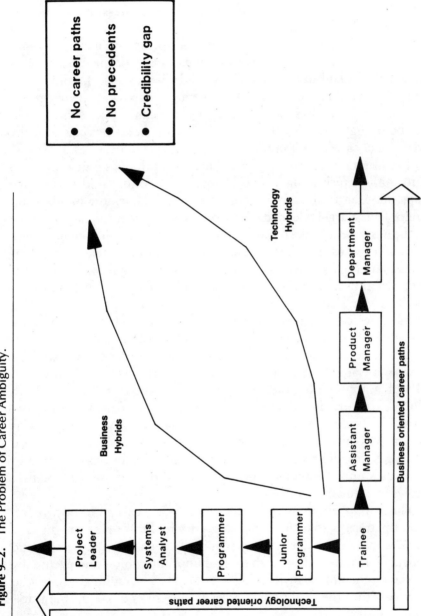

and no longer being part of the mainstream of IS. At the same time he or she is not a "real" finance or marketing professional, however many systems the information center helps the finance and marketing functions build. In the same way the junior person in finance who has learned Lotus 1-2-3, FOCUS, Multiplan, or any of the other myriad new tools for end-user development, and who now works almost full time on building applications using them, is not a real systems professional, and in fact has only a smattering of technical skills that would not be enough to get a job in IS. He is also no longer moving along the traditional career path in finance.

The situation is extremely ambiguous on both sides. Integrating the information technologies and bridging the culture and knowledge gaps between technical and business people at senior, middle, and junior levels largely depend on maximizing career ambiguity for some of the best talent in the organization. Just a few examples of new roles without clear career directions are office technology, telecommunications planning for business innovation, supporting personal computing, defining information needs for customer service, developing products for electronic delivery, financial planning for IIT, training users of personal computers, developing strategic plans for computer integrated manufacturing, and selling such technology-based products as electronic cash management and dealer order entry systems.

In some large firms, senior management's response to the problem of ambiguity is to tell the people concerned about the issue: "Don't worry—you are the ones who will be running the firm 10 years from now." They *do* worry and they need an environment where what is promised is not a punishment, but a privilege. Systematic lateral assignments from business groups into IS for periods of six months to two years must become an institutionalized part of the way the firm grows its human resource.

This is happening in a very few companies, except ad hoc. A frequent pattern is that the junior analyst who builds planning models for the marketing department is initially seen as Superperson and quickly becomes indispensable—and unpromotable—and eventually moves on to another firm to do the same job. The junior analysts who stayed within the traditional career frameworks move up the traditional ladder.

The hybrids are the new blood of the organization of the 1990s. For them to grow, they must have management attention, if only because in a time of ambiguity people will watch what management does, not what it says. There are many formal and informal ways senior and middle-level executives can help:

- Selecting the best junior staff for the lateral development that builds hybrid skills
- Redefining the management "fast track" to require experience in taking full and direct responsibility for some aspect of IIT
- Helping break down the physical and psychological isolation of the IS organization by insisting that staff be temporarily assigned to work in business units
- Providing first-rate instead of expendable supervisors to provide user involvement on development projects

All these are relatively simple contributions they can make.

ORGANIZING THE INTEGRATED INFORMATION FUNCTION

The roles needed for managing the integrated technologies are changing and so too must the organization that has responsibility for them. In most firms, that consists of central information systems function plus divisional IS units, and a corporate telecommunications group, and any number of fragmented units that handle some aspect of voice and data operations. The function of the telecommunications group largely depends on the stage the company is in, in the shift from a technical utility to a coordinated business resource.

The specific mandate, structure, and relationship of IS to senior management and to the wider organization largely reflect the historical development of data processing and of the communications utility. Very rarely are these a conscious response to the realities of the electronic marketplace. A new approach to organizing IS is needed.

Organizing is not the same as organization. It is easy to draw new organization charts and shuffle jobs around. "Organizing" implies a much more dynamic emphasis on communicating, and on roles—literally, the parts they play—rather than tasks. On the whole, telecommunications and information systems organizations have been defined more in terms of tasks than roles: projects, specialist skills, technical niches, and responsibility for specific applications. The key themes in organizing IS have related to building systems and running corporations.

The new roles related far more to marketing, communicating, supporting, and planning. The old skills remain essential. If anything, solid, reliable operations are more not less important when failures affect service and are seen by customers instead of being hidden behind the walls of the data center. But the new information function is a full-service company, rather than a manufacturing department.

This means that the criteria for organizing become

- Coordinate the planning, implementation, and use of the information resource, balancing central direction with decentralized application.

- Shift the organization for IIT toward being a staff function comparable to corporate finance, instead of being mainly a unit that builds systems.

- Amalgamate telecommunications and information systems within the IIT organization as interrelated departments within the information company, not independent functions.

- Use human resource planning to drive, not follow, technical planning and implementation.

Figure 9–3 shows the likely formal organization structure that results from these principles.

Suitable Centralization: Coordination, Not Control

No organization chart can communicate the main role of the IIT function: to find a suitable level of centralization to coordinate the integrated business resource. The main principle for managers of IIT has to be: Respect the reality of decentralization and establish the criticality of coordination. Very roughly, this means that the key infrastructures, especially telecommunications, require a move toward centralized direction and that building the applications that carry the traffic and making decisions on what traffic to add should be increasingly decentralized. If the architecture is clear and technical standards are backed up by rules, guidelines, and procedures, "distributed" IS units in the business groups can handle most development needs. This is a very big "IF."

Figure 9–3. Degrees of Centralization in Managing Telecommunications.

	Strategy	Role of Corporate Telecommunications Group	Standards	Priority	Organizational Pros and Cons
Local Autonomy	• Business units take full responsibility for planning and operations	• Advisory only; may coordinate investment plans	• Minimal, if any	• Cost-efficient local operations	• Matches planning and operations to decentralized units' very different needs and environment • Guarantees incompatibility • No basis for gaining economies of sharing
Guided	• Local autonomy plus central unit to provide technical and planning assistance	• Provides guidelines and consultancy services, especially in areas involving new technology	• Standards are recommendations and mainly based on technical and cost issues	• Cost-efficient local operations plus providing economies of expertise	• Builds small, highly skilled advisory group local units can draw on • Standards often ignored since authority lies with local units
Coordinated	• Central group plays strong staff role and sets criteria for long-term planning and investment	• Coordinates planning and may supervise operations of some corporate facilities	• Guidelines with teeth; corporate group has some authority to ensure standards are followed and approve exceptions	• Balance between centralized coordination and decentralized development and operations • Creation of flexible architecture to provide integration path	• Provides base for corporate highways with local freedom about traffic • Frequent ambiguities, or conflicts, of authority
Centralized	• Creation of a corporation-wide set of shared utilities and common delivery base	• Directs planning and operations • Ensures standards are followed; may leave day-to-day operations to local units	• The standards are rules to be followed	• Corporate rather than local cost efficiency • Creation of shared corporate resource	• Provides clear links between corporate business planning and telecommunications planning • Frequent opposition by local units to central "czar"

Guidelines with Teeth

B. Williams of the consulting firm Arthur D. Little provides a useful framework for assessing the degree of centralization of any particular telecommunications organization. He defines a spectrum ranging from fully autonomous to integrated/centralized. Each extreme seems undesirable. Autonomy, with every unit making its own decisions, means that a corporate resource as opposed to a set of incompatible technical facilities will never be created; full centralization contradicts the principle of decentralized decision-making that is one of the most basic realities of modern business.

The two intermediate positions, "guided" and "coordinated," are unstable—they combine a little of each extreme. They both have many merits, but they can be made to work only if senior management defines a vision to justify them, establishes the policies to make them possible, and clarifies the authority and roles of the centralized architect and planning unit and of the decentralized business operations. The spectrum of organizational choices and some of their implications are shown in Figure 9–4.

The phrase "guidelines with teeth" addresses the issue of authority. "Standards" is an ambiguous term. It can mean anything from a set of recommendations that can be ignored by business units to corporate architecture where there is no local discretion allowed. The technical standards may be the same in both instances; the key question is who either enforces them or decides on exceptions?

As one telecommunications manager in a major international consumer goods firm discovered, publishing a set of standards does not answer that question. His company, Quintex Inc. (a pseudonym), planned to install a common worldwide financial reporting system; to support this, the manager defined a set of computer and communications standards to be followed by each division in each country in which Quintex operated.

The head of Taiwan subsidiary decided not to wait for the new system. The volume of business was expanding rapidly and existing computer systems were inadequate to handle the unit's needs. He approved the purchase of a software package that ran on a "nonstandard" computer and used the equipment manufacturer's nonstandard communications protocols.

The corporate telecommunications manager tried to get the decision overridden. He pointed out that he had been given the responsi-

Figure 9–4. Organizing for the Electronic Marketplace.

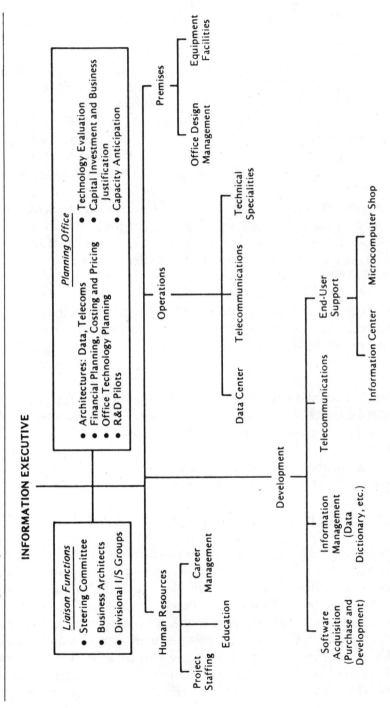

Key changes to existing Information Systems and Telecommunications organization:

- Bring together telecommunications and systems development
- Create a strong planning office
- Separate development and operations
- Build strong human resource function

bility to define a companywide policy and that Taiwan's action threatened the whole concept of shared computer systems, common reporting, and integrated communications. The head of the subsidiary responded in a memorable telex that he had been given the responsibility to make profits for the firm, that he had immediate needs and did not intend to sit around and wait.

The communications manager responded that the local decision threatened long-term integration. He argued that it would add costs in the end and slow down the implementation of the worldwide system. He lost the argument and Taiwan has its independent communications facility.

It is not easy in such instances to decide who is right. Here, it did not matter; the senior business manager had the authority and the communications manager had nothing more than a piece of paper and might was right. For standards to be effective, they have to have teeth. The communications manager recognized this after the event. He asked the information systems steering committee, which set policy for computing and communications, to clarify his mandate.

He listed the benefits from the integrated approach and the costs of local autonomy. He accepted that sometimes exceptions would have to be made but argued that he must have, if not the final say, then at least some real influence. He wanted "a preliminary veto." All local systems must follow the standards, and his unit's approval should be required for any deviation from the standard. He would have vetoed the Taiwan proposal and suggested that they find an alternative package. If Taiwan wanted the veto overturned, they would have to make their case to the IS steering committee. "The onus should be on the divisions to justify exceptions, not on me to justify the standards."

THE CORPORATE ARCHITECT

Most large firms have an architecture or are trying to define one. That is a basic requirement for any coordinated communications plan. They need to make sure they clarify the role of the corporate architect. At the extreme of fully autonomous management, he or she is a staff adviser and at the other extreme, fully integrated and centralized, a controller. Guided or coordinated management requires a custodian—someone who can guarantee the integrity of the overall architecture, while adapting it to special needs and exceptional situations. This means having at least a preliminary veto. Quintex's manager defined an architecture, but his role as an architect was not made clear.

In almost every firm where telecommunications is seen as more than an internal technical utility, the trend is toward centralization. This assertion is based on analysis of almost 50 large companies, half of which are U.S. companies and half European. They come from every major industry. The analysis is based on published and private documents and was carried out in mid-1984. Follow-up analysis in late 1987 confirms the trends; the only main change has been that telecommunications is more and more interlinked with information systems. Regardless of their specific strategies, almost every firm is trying to resolve the issue of how to get centralized coordination.

None wants to push toward more decentralization of planning, although most of them favor decentralized network operations. The firms with central units most successful politically and technically share the following characteristics:

- The central group plays a strong consultancy role for the other units. It relies on having a pool of first-rate technical staff to encourage them to draw on it; rather than try to control by fiat, they control by incentives and expertise.

- They have a clear mandate to design and maintain the optimal topology for the corporate network, supervising almost every aspect of planning and operating facilities that are identified as "corporate."

- They use accounting mechanisms and provision of service and support functions to ensure credibility, equity, and incentives for their client units.

The accounting mechanisms include charging by use, not budget, with a predictable standard cost per unit, such as a transaction or monthly subscription fee per terminal. Another mechanism is providing both consolidated and detailed billing reports so that user managers can predict and control their usage. Many of the firms try to set their prices below the outside market rate.

In many of the firms, especially outside the United States, relocation of operations motivates coordination. Many of them are moving information systems and administrative units away from central cities. Another trigger is office technology, and a more general concern to reduce overhead and staff costs. A third force toward coordination is simply underestimating growth in demand for telecommunications. One oil company predicted and planned for a 20 percent increase per year in the number of terminals in use between 1980 and 1985; the actual figure was 70 percent. Bank of America anticipated in 1982 that it would process 80 transactions a second on its main online systems by 1986. In 1984 it revised the estimated to 500 to 1,000. A British insurance firm based its five-year plan for 1986 on the assumption that it would have 1 workstation for every 10 workers by then. By late 1985, the ratio was already 1 to 3.

THE INFORMATION EXECUTIVE

The title "chief information officer" (CIO) has recently become popular in describing this new style of business-oriented executive in a previously largely technical role. One natural question is: Does the mandate create the person or the person the mandate? Is this a job title or a way of operating? Just relabeling the old-line data processing manager as a CIO does not change his or her skills, attitudes, and ways of behaving. The information executive needs some distinctive personal skills, especially the ability to communicate in the widest meaning of the word, to translate between the worlds of business and technology, and to earn attention from and credibility with senior managers. Information systems and telecommunications managers need to be honest with themselves in answering the question "Why should the CEO give me the mandate and title?"

The CEO should also be honest in answering a parallel question "Is top management the blockage to creating an effective information executive role?" Figure 9–5 summarizes the managerial resources and style of operating that facilitate or inhibit its emergence. Figure 9–5 (a) shows four somewhat stereotypical types of information systems and telecommunications managers.

The Monopolist is the old-line technocrat, who had complete control over the corporate computing resource, and resisted end-user computing and personal computers; monopolists have (or perhaps more accurately, had) plenty of authority and little if any responsibility for business development, support, and innovation.

The Information Janitor has neither the authority nor the business responsibility to be anything much more than an operator of an efficient data center; the action is elsewhere, often in the business units. The Whipping Post has—or tries to take on—a high level of business

Figure 9–5.   IS Leadership Styles and Constraints.

(a)   Authority and business responsibility

The IS leader's level of business responsibility

The IS leader's level of authority

	Low	High
High	WHIPPING POST	INFORMATION EXECUTIVE
Low	INFORMATION JANITOR	MONOPOLIST

(b)   Style of the business versus the IS leadership style

IS leadership style

The business leadership style

	Passive, reactive	Active, innovative
Active, innovative	MISSIONARY	INFORMATION EXECUTIVE
Passive, reactive	LOSER	FOOTDRAGGER

(c)   IS management styles

IS leadership

Business leadership

	Passive, gives IS limited authority	Active, gives IS real authority
Active, takes on strong business responsibility	MISSIONARY/ WHIPPING POST	INFORMATION EXECUTIVE
Passive, avoids or ignores business responsibility	LOSER/INFORMATION JANITOR	FOOTDRAGGER/ MONOPOLIST

responsibility but lacks authority. The Information Executive combines both authority and responsibility.

Figure 9–5 (b) complements the first one. It shows the impact of passive and reactive versus active and innovative leadership at the top of the business and the top of information services.

The Loser is the passive, reactive IS manager supporting a business leadership that is similarly passive and reactive in its views of the use of integrated information technologies for competitive positioning. The Footdragger, who is often also the Monopolist, has fallen behind the business leadership and plods onward, taking the attitudes and techniques of the past into a future where they are less and less applicable. The Missionary is ahead of the business. The Information Executive is an activist supporting and supported by activist business leaders.

The two diagrams can be combined (Figure 9–5 (c)). Losers/Information Janitors are largely irrelevant to business in the coming decade. The Footdraggers/Monopolists are part of the past that can slow down the future. The Missionary/Whipping Posts are the people who—given improvements in business management awareness, changes in policy, and understanding of the competitive urgency of IIT—can become Information Executives. Without these, they generally become consultants.

The Chief Information Officer/Information Executive is created by a combination of IS leadership, personality and style, business management leadership, policy, and real commitment. In 1978 information systems and telecommunications units in the Fortune 100 were largely populated by the footdraggers and monopolists. In 1988 there are more and more missionaries looking to create the dialogue with business leaders that multiplies the ability of both to be practical futurists.

LIAISON FUNCTIONS

Some of the mechanisms for two-way coordination include steering committees, liaison units for applications, and formal and informal planning groups that include representatives from the business units, divisional IS group, and all other groups affected. Their job is to establish priorities and plans, and to resolve issues of resource allocation, phasing, staffing, and project management.

One key liaison mechanism requires much senior management direction or attention, especially when the firm is trying to accelerate its use of technology to make major business moves. That is the top policy committee of managers who can make things happen, and who share, review, and update the business vision for telecommunications and make it credible. This is not just another committee. If it includes two or three of the most senior managers in charge of key parts of the business, a few who may be more junior but are recognized in the firm as innovators and opinion leaders, plus the head of any corporate function who is a very powerful force in the firm, then the title of the committee is irrelevant. Citibank called its planning committee for international electronic banking "the gang of seven." Tom Theobald, its head, assigned some strong personalities from marketing to direct its efforts. One of these was famous for his aggressive money making for the bank. The technical organization in the international bank was very weak, and no bland committee could have given it the political muscle that the gang provided. The selection of these people sent a clear signal to the rest of the firm of the importance the bank placed on telecommunications.

CROSS-FERTILIZING THE PLANNING OFFICE

No firm that is trying to move quickly to exploit IIT has the skills in-house to direct its planning. IS and TC leaders need accountants to find ways of funding, costing, and pricing what is now a complex economic good. They need marketing experts to shift to a service role. The information center, support for end-user computing, assignment of client or account managers to handle particular business units' needs, and the development of business systems analysts are only a few areas where information systems and telecommunications groups are adopting a market-oriented rather than product- and application-based service role. But they have very little knowledge of marketing principles and techniques.

In the same way, IIT needs economists to help in forecasting, pricing, and modeling the relationship between the business plans, the external environment, vendor strategies, and trends (especially IBM products and prices and telecommunications costs) and the technology strategy. It increasingly has to be able to draw on people who understand ergonomics, psychology, education, and human resource planning because every aspect of IIT is now intrinsically interlinked to behavioral and organizational issues.

Sincerity is no substitute for technique. The best IIT organizations are sincerely doing their best to broaden their planning base. But sincerity just isn't enough. Techniques come from:

- Systematic cross-fertilization between the IIT unit and the wider organization, by recruitment and temporary assignments from within the organization, and by distributing development and support staff to the business units

- Using consultants and academics to bring in new knowledge and avoid personal obsolescence, to provide an integrating perspective, and to help educate and advise senior managers

- Hiring, growing, and retaining good talent

Figure 9–3 lists a few of the planning areas for IIT that have to be addressed by some combination of these three sources of talent. Obviously, it is up the IIT organization to handle most of these aspects in the short term the best way it can and by human resource planning for the long term.

To encourage staff to seek and accept the temporary assignments to or from IIT necessary for cross-fertilization, present such jobs as a privilege accorded to the new "fast track." Reassure the assignees either by guaranteeing a position to return to or by creating permanent liaison roles. Insist that there be no poaching. Many IIT units report that their users, quite sensibly, try to retain any good people who are loaned to them. Remember that the goal is to build a cadre of hybrids. Time the lateral development carefully: It is silly to take someone who has 20 years' technical experience and very limited exposure to the business and expect him or her to be credible or effective outside that sphere; the VP of Systems Development should not become an overpaid apprentice in marketing.

Conversely, if trainee programmers are shipped out to Finance, they are of little value since they have not yet mastered their own trade. The best time to make the move seems to be between two and five years after a person's employment. The move across should last no more than two years so that the person's old knowledge does not become obsolete.

If 10 percent of incoming recruits are targeted for such a process in both directions, the long-term human resource problems for IIT will be solved in about five years. The sooner a firm starts, the quicker the chief bottleneck to exploiting the business opportunities of technology will be unjammed. Good IIT managers and good personnel managers know this. They generally cannot do much to solve it, because it requires both senior management initiatives and clear signals from the top that this is the path for the future.

As for consultants, in the field there are bad and good consultants, greedy and responsible ones, merchants of hype and solid professionals. It can be hard to tell which is which, especially in new technical areas. The best advisers often apply their business experience to a new area involving technology although they will not be able to show success in it.

Often senior executives bring in academics or well-known experts to help them get a sense of the main options and issues. When they do, however, managers should beware of joining the fad-of-the-month club. When *Business Week* in 1985 published a cover page article on the coming wonders of expert systems, for example, many IIT managers groaned. The article, like so many on a hot new topic in the field, gave managers false expectations about the progress in expert systems and artificial intelligence (AI). It led to their pushing for action and bringing in experts to talk about what the firm should be doing.

These people often come from a scientific or academic environment where their experience is with small-scale applications or pilot projects; that is where work at the state-of-the-art is being done. They do not know what they do not know. They have no understanding of large-scale commercial processing and operations, of organizational aspects of information management, of the difference between feasible and practicable. It is not that their knowledge is invalid, only their extension of their knowledge to the world of business and business uses of technology.

The main differences between the worlds of academic/scientific and large-scale business computing are shown in the following list.

Scientific/Academic	*Commercial*
Analysis is complex in methods, models, computation.	Operation is complex in terms of project management, coordination, procedures.

Data structures are complex, volumes low; advanced software.	Volumes limit complexity of data and practicality of software with high overhead and inefficiency.
Technical base is stand-alone or simple time-sharing computer.	Operating system and architecture dominate decisions.
"Architecture" refers to the hardware.	Architecture relates to the integration of the organizational resource.
The organizational context is of limited relevance to technical decisions.	Organizational interactions are driving factors for decisions.
Implementation is equivalent to installation.	Implementation means making the system work organizationally as well as technically.
The application is the strategy.	Business operations define the strategy.

It's Easy.   Bright people who operate only from experience and assumptions often make the IIT manager look like an anti-intellectual neanderthal; he or she is pushed onto the defensive. Until very recently, the field of information technology in large organizations had substantial barriers to entry: experience in the trenches of systems development or telecommunications operations. That barrier has been removed. The areas where firms most often need top level advice are ones in which such experience is not necessarily relevant. Business planning for telecommunications is a particular one. The best advisers will be people who are thoroughly up to date on developments in Open Systems Interconnection (OSI), advances in local area networks, value-added networks (VANs), ISDN, or digital PBX. Examples where, because a concept is proven or a framework available, the inexperienced adviser says it's easy include large-scale database management, international standards for telecommunications, implementing local area networks, and voice/data integration. It is not easy and it is misleading to persuade managers it is.

Have Transparencies Will Travel.   The IIT field is grasping for ways of making sense out of the bewildering changes and dilemmas firms face. But they need simple, not simplistic models. How do they tell the difference? The deep and the shallow thinkers use the same transparencies in their presentations. Conceptual models and mere notions look and sound the same. The deep thinkers have a base of proven results, either their own experience or the successful implementation of their recommendations.

Have Good Idea, Will Sell. Sometimes, a leading adviser, especially on the busy conference circuit, comes up with a useful framework or striking message. That speaker will be in constant demand. The temptation is then to stop learning and to stick with the one idea, well and often wittily presented. Again and again in IIT, good academics and consultants have used up their intellectual capital by turning it into temporary income.

IBM-Phobia. This is a problem of intellectual adolescence. IBM represents the orthodoxy of information technology. Many of the Young Turks in the consulting field are explicitly anti-IBM.

There is a lot of truth in the criticism of IBM (see Chapter 7), but the new IBM is not the old one. It has been the aggressor, not the follower, in the market. With all its faults, it has moved to head off all its competitors, including even AT&T, except the Japanese—this now is the battleground. Most important, it has established its architectures, rather than its products, as the reference for the field. SNA is the de facto standard. That is why every major manufacturer of computer and telecommunications equipment has adopted it. Its personal computer has never been the best in the field, nor have its office technology products generally matched the best of other vendors.

Experienced IIT managers and planners know all the pluses and minuses of IBM, from the value of the plastic wrapping to the cost of its cumbersome operating systems. For them, the issue is "If not IBM, then who?" For the new generation of consultants, the issue has tended to be "Not IBM." As a knee-jerk reaction, that is as silly as the stereotypical old-line data processing manager's on-the-knees reaction of "Only IBM, of course."

QUESTIONS TO ASK WHEN CHOOSING AN ADVISER

Does the prospective adviser understand the craft of large-scale information systems and telecommunications development and operations? Has he or she worked on a really big project (involving at least a dozen people for a two-year period, say) from inception through to operational use? If not, there is a fairly strong risk that he does not know what he does not know and overlooks the complexity of the management process and the interaction between technical and organizational issues in large-scale business applications. One cannot simply extrapolate from scientific projects that involve small, simple or well-structured databases, or from personal computers and local area networks, to an environment marked by a myriad of interdependencies.

What is the candidate's pedigree, in terms of intellectual base and the quality of the firm(s) he or she has worked in? The exemplar firms described in Part II provide their people with a training ground that in itself adds a value to their own qualifications, job titles, and project experience. The same is true for the very best of the computer and communication vendors.

Does the adviser know the field, in terms of the research literature, what is going on in the leading vendor and user companies, and the practical state-of-the-art? Ironically, at a time when the field of IIT needs a combination of first-rate analytic ability and some mix of breadth and depth in business and technical areas, anyone can become an "expert." It is very easy for people to grab the latest fad term or hot topic and sound convincing. This happened in the early 1980s with office automation, and more recently with expert systems. It is increasingly commonplace with business telecommunications, which is among the hottest of all topics.

The reason senior managers have to address the issue of how to validate outside advisers for IIT is simply that they have no choice but to use them. The same is true within the IIT organization. Bad advisers do damage, however. They raise expectations about what is practical, mislead the organization about the risks and returns, and add to fog and fantasy, not vision.

Development and Organizations

The most immediate change large firms have to make in how they organize the IIT unit is to move from having information systems and telecommunications as separate functions, each with a development and operations unit, to splitting the IIT organization into a development arm and an operations arm, each of which includes telecommunications and information systems groups within it.

In many large firms telecommunications and information systems have evolved on largely separate paths. In addition, the pace of change in communications, from voice and analog to digital technology, and in information systems from automation of clerical processes and "batch" systems to a broad range of on-line and data- and communications-oriented applications, have both increased the fragmentation of responsibility and authority. The company then has voice specialists who do not understand data communications, data communications staff who disdain data processing as a technically unsophisticated function, and data processing specialists who view telecommunications in terms of the software requirements for on-line applications. The development staff have up-to-date technical knowledge and relatively little experience, while the operations personnel have the solid experience and an obsolescent knowledge base.

The cultures have to be brought together and the organization based on each one's strengths. In the development arm, the information systems functions include

- *Software acquisition to build and buy applications.* The trend is toward buying packages, "end-user" software, and "fourth generation" languages. Standards for system compatibility are important.

- *Information management.* Data is among the most valuable traffic on the network highway, and the delivery of and access to data via telecommunications is a competitive resource. Database management software and procedures and technology to create the data architecture shift the focus in IS from programming to information resource management.

- *End-user support.* An information center should be a do-it-yourself store for nontechnical people to develop ad hoc small-scale systems, and should offer assistance in acquiring and using microcomputers.

There is a growing trend in leading firms to create an account management function for IIT. Instead of the traditional project-based structure, they operate as a service unit either analogous to IBM or comparable leading vendors, or a joint venture for partnership—a word that appears in more and more statements of the IIT unit's mission.

First-rate operations remain critical. Reliability, security, response time, smooth installation, maintenance, and troubleshooting translate to quality of customer service in the business resource era. When the network is down, the business is down. The old-timers are not obsolete. They provide a critical skill. Rather than turn them into mediocre development supervisors (and the digital communications whiz kids into ineffectual managers), surely it makes most sense to recognize that strategy needs cables as well as vice versa.

LAST BUT NOT LEAST: FACILITIES

The final major innovation in organizing for IIT is one whose importance is easy to overlook: the fact that the office of the future is very physical in nature and telecommunications has tremendous implications for office design and administration, and vice versa.

Many telecommunications-related functions are controlled by administrative service units, especially at the local level. When an employee now moves to a new office, there can be several days of work needed for drilling and cabling to install a workstation. Furniture, lighting, and desks have to be picked on the basis of ergonomics and health and safety factors for users of terminals. No office can be designed now without careful attention to cabling for local area networks, PBX, and a ratio of employees to workstations that is virtually certain to move close to 1:1 during the lifetime of the building.

The information executive needs new authority over many aspects of office design, must supervise any relocation of business operations, and control the planning of office equipment and facilities. This is a very different aspect of IIT from development and technical operations.

Facilities management has traditionally been a subset of administrative services. The move is now in the other direction. One British manager of IS in a large insurance firm is quite candid about his objective here: "I want to get full control over Administrative Services because by 1990 the main evaluation criteria for my group's performance will depend on trivial details that they now handle."

10 MOBILIZING FOR RADICAL CHANGE

Most of this book has been about senior managers' responsibility and opportunity to use telecommunications to help ensure their firms' economic health over the coming decade. They also have a responsibility for its organizational health too. Telecommunications and organizational change can be smooth or disruptive.

Senior managers will determine—sometimes by default—which is the more likely case. There are many examples where they ignore or are even indifferent to three factors that explain many of the organizational failures of technically successful innovations:

- *Time* is the component of information technology that the vendor's ads never mention. Unrealistic deadlines are disruptive.

- *The need to view technical change in organizational terms* and to provide resources for managing it, including the key directing influence of their own time, prestige, and attention.

- Education as a mobilizing force to prepare people for change and help them participate effectively and adapt easily.

In this chapter are just a few examples of where vision, policy, and architecture have not been complemented by a suitable organizational strategy for mobilizing. In several of the cases, the cost was mainly economic—the firm was not able to get the full benefits of its investment. But in others the damage went further: people were hurt badly and unnecessarily. In all the instances, top management was the main cause of the problem and could and should have been the main cause of making sure it did not occur.

231

THE IMPOSSIBLE DEADLINE, AND
OTHER MISTAKES TO AVOID

In a leading international bank—let's call it Megabank—senior management authorized a drive to improve back-office efficiency and installed a complex new processing system. The technical planners estimated that it would take several years to implement. The manager brought in to spearhead the project insisted that it be done in nine months and made it very clear that heads would roll if there was any delay.

Progress was initially fast but many, many problems occurred later on. The progress was in fact illusory. People lied. Management by fear led to them having to do the impossible. They knew they were in trouble if things were late, so they pretended they were on schedule and hoped they could catch up as they went along.

They put pressure on the back-office staff, who had to keep the old system going and install the new one. The stress was appalling, with burnout and breakdowns. There was even a threatened sick-out by workers. Management's response was to set even tighter deadlines and exhort supervisors to push their staff harder. But the project kept slipping. There was not enough time for any amount of effort to resolve the problems of change in work, procedures, skills, and technology.

Frederick Brooks, whose book *The Mythical Man Month* explains why estimates of the time needed to implement a computer system are so often underestimated, summarizes why time is so often an immovable constraint in large-scale projects: "When a task cannot be partitioned because of its sequential steps, the application of more effort has no effect on the schedule. The bearing of a child takes nine months, no matter how many women are assigned" (page 17).

In the case of Megabank, senior managers demanded a three-month baby. It was not they who had to pick up the pieces or handle the strain. They put the pressure on because the firm was in serious business difficulties. When firms have to move fast in telecommunications, because of competitive pressures, it is natural for senior managers to set deadlines based on what they want to have happen. With integrated information technology, that often does not work.

Building a major new system or making changes requires from two to five years within the existing architecture; business innovations that involve changing the architecture take close to seven years, and though it may take only one year to install a system, it takes two or four more before it works organizationally. New systems cannot be parachuted in. They need to be assimilated. Education helps speed up the process and the technical quality of the systems design and the effectiveness of project management can also accelerate progress, but the baby takes nine months from conception to birth.

Many planned IIT projects fail right at the start, when senior management specifies an operational date of July 1. Some of the most publicized "successes" of business innovation are in fact failures—installed but not really implemented, technically a success but at the expense of a demoralized work force and burnt out implementation teams, and unnecessarily expensive when all the costs, economic and organizational, are counted. The implementers publish an article in 198X about how they have reorganized work and dramatically improved productivity. In 198X + 3, it is clear the victory was illusory. The company is still trying to make the installed system effective.

THE VISION IS NOT CONTAGIOUS

Senior management at Gigawidget (this is a pseudonym for one of the top five automotive parts makers) committed to a bold plan to improve the quality of and ease of access to information across the firm. The plan would have a substantial impact on long-term efficiency and allow business units to manage their marketing and production more effectively.

Unfortunately, the units had many short-term needs, especially for ways to cut costs. The new management information systems (MIS) project rested on defining a data architecture and a communications architecture. The lead time was three years for the initial applications, with an open-ended timetable for eventual completion. The business justification rested on top management's assessment of coming major shifts in its manufacturing sector.

About two years into the project, there was a backlash from the business units. They became increasingly critical of the decision to invest in a large computer mainframe and to upgrade corporate communications facilities. The senior vice president who had been the key supporter of the project summarized what had happened:

We overestimated the extent to which the divisions shared our view of the priorities for the firm. They got frustrated because they have immediate needs for MIS and many of these can be met by personal computers. They ask why we have to wait three years to get results. They don't see a need for an architecture. They do see a lot of money being committed.

In retrospect, it is clear that we have failed to move them with us. We should have invested in education early, so that they could understand the business issues, why we have to have an architecture and why it all takes so long. We didn't do this and I'm afraid we may have lost credibility and the time window for action has closed on us.

This is a company where top management sees integrated information technology—electronic coordination and customer service plus management information—as a key to its survival. The rest of the firm is as aware as the people at the top that business conditions are changing. They also view information technology as a necessity and a force for innovation. In no way are they managerial Luddites.

They do not understand vision, policy, and architecture. Given the pressures they face and their limited understanding of the management issues for IIT, their opposition to the MIS project is not surprising. They want departmental systems now, not corporate ones sometime.

Education could have helped build awareness and change attitudes. The senior VP of Gigawidget sees it now as "Absolutely essential. We should have done it right at the start."

The firm had a plan for technical and business change, but none for handling organizational change. The new failures in introducing IIT into the workplace will come from neglecting the realities of time, the need for education, and the importance of recognizing the organizational side of innovation rather than from flaws in the technology itself. The failures will be very expensive, since they will affect the firm's competitive positioning, not just its operational efficiency.

SAVE NOW, PAY LATER

A leading computer manufacturer spent over $18 million on a new competitive database, COMIS, short for "Competitive Information System," which included estimates of key customers' expenditures on individual vendors' equipment and new product announcements. It encouraged marketing staff to access the system.

They did not do so, even though the marketing planning group sponsored a number of half-day seminars on how to use COMIS.

They had interviewed product managers on their information needs; up-to-date competitive intelligence was uniformly viewed as a priority.

Unfortunately, the marketing planning department provided training, but not education. The seminars showed users how to create a COMIS report to produce graphs and even set up a private database for off-line analysis on the personal computers.

Barbara Grace, an IBM researcher, anticipated the difference between training and education in 1975. She pointed out that non-technical users of computers have to climb three steps to become comfortable, skilled experts.

1. Training: How do I use the system?
2. Support: How do I use it in my own work?
3. Education: How do I use it my way, in my job?

These training seminars answered those questions and showed the product managers how to use the technical system. After that, they could go home and demonstrate to their children how to log on to COMIS, produce a pie-chart, or find out which vendor sold the most personal computers to General Motors (the answer was IBM, which the managers already knew).

The training program left the users in the situation of a British judge who commented to Norman Birkett, one of the most articulate of all Britain's verbal gymnasts: "Mr. Birkett, this leaves me none the wiser." "No, your Lordship, but infinitely better informed."

The informed manager needed to know how to use COMIS in his or her own job. "What data will help me direct my sales force to identify the main issues customer X is concerned with? How do I identify which competitors are making inroads on my key accounts?" Answering these questions requires an internal consultant to support users and to help them build ad hoc models and reports.

Even this does not address the question "How do I use COMIS in my own way? I have a very different market to deal with from most of the product managers. I have to be able to answer sales representatives' urgent queries like "It looks like Ford is going to issue an RFP for its new dealer system. DEC and HP will probably bid. Can you help me scope what they are likely to offer? I'm told COMIS can do this, but no one in MP (Marketing Planning) seems to know if that's so."

The response from the designers of COMIS is "Jim keeps complaining, but COMIS has all the data he needs. I don't have time to keep answering his questions. Why doesn't he spend time learning how to use COMIS properly?"

Jim is well trained and undereducated about COMIS, which is yet another example of a first-rate technical design that is rapidly falling into disuse. The responsibility may not at first sight have anything to do with top management, but. . . .

> Our initial proposal for COMIS included assigning two people as full-time business consultants to the product managers (PM's). We understood very well that they are not used to operating a terminal and, quite honestly, don't know how to analyze data, even though they'd deny that.
>
> The budget was cut back. It's head count cutting time again. We can't make the case. Bernie (the head of Marketing) is willing to spend $200,000 on the system but won't—or can't—add a $20k analyst.

The product managers are indeed infinitely better informed but none the wiser, and they do not use COMIS.

Megabank, Gigawidget, and COMIS are not exceptional examples. Time, education, and resources for helping people adapt to change are intrinsic requirements for making new systems work organizationally. TINFL (There is no free lunch).

Dear Senior Manager—Save Now, Pay Later. You cannot get the benefits of IIT cheaply; of course you can cut the budget for education and set arbitrary deadlines. BTWW (But that won't work). AIND (And it never did).

To institutionalize IIT successfully, the firm has to provide three types of resource, all of which are expensive: technical, managerial, and organizational.

LESSONS LEARNED ABOUT MANAGING TECHNICAL CHANGE

An extensive literature built up over the past 25 years explains the reasons for success and failure in introducing information technology into the organization. There have been plenty of failures to study: in the late 1970s, it seemed as if "resistance to change" were the main explanation—or excuse—for the gap between management's expectations and designers' intentions on the one hand, and ineffective implementation and/or use on the other hand. The conclusions from all the studies are clear-cut. What is needed in introducing technology is

- Evolutionary rather than revolutionary phasing in implementation
- A great deal of organizational resources and effort (A rough rule of thumb is that training alone should take up 20 percent of the budget.)
- Sensitivity to internal politics
- Face-to-face facilitation between the technical staff and the users
- Small-scale pretesting and piloting
- Attention to the qualms of resisters, who may be reasonable people who see the cost of change as greater than the benefits

These points contain a paradox: Effective implementation of technical change relies on evolutionary change, but telecommunications aims at radical change. Twenty-five years of trial and error in data processing and management science, backed by a wealth of studies in political science and sociology, warn us that successes here are rare and failures commonplace.

Worse, now that telecommunications is being used as the enabler for customer service, market innovation, internal communications, and managerial and administrative productivity, change is the norm. The new cliché for integrated information technology is that the only constant is change.

None of the problems of change should discourage firms from moving fast in telecommunications. Instead, they indicated the opportunities for the firm that combines vision, policy, and architecture with a coherent plan for mobilizing the organization. They give the edge to the company that can move its culture effectively, since most firms will not be able to take on radical change.

Leadership and "Culture"

The term "culture" like "excellence," "vision," and, more recently, "entrepreneurship" has crept into the business vocabulary within the past few years and it is worth asking why. Is it because in a time of growing strain and uncertainty, organizations have a sense that there is something intangible, a mixture of morale, attitudes, shared values, and expectations, that has to be preserved for the firm to be able to steer a difficult course? Whatever "it" is, there is something special that has to be added to the recipe for business change.

A striking feature of many of the firms that have been most success-ful in innovations in the electronic marketplace has been their ability to move their culture with them; while there must have been some degree of resistance, they seem to be able to institutionalize the idea of changing. Dun and Bradstreet did this through planned programs of support and education. American Airlines, under its CEO Robert Crandall, Citibank under Walter Wriston, Scandinavian Airlines led by Jan Carlson, are firms where the leaders have made their own style and personality almost an organizational value. IBM made just about every mistake possible in reading the market in the late 1970s but the culture remained vital and when they were given a new signal that it was now permitted to move, IBM's people did so aggressively and well.

There are many other firms that somehow have not been able to mobilize their culture in the same way. Chase's strategies for elec-tronic banking and the technical qualify of its systems have often been way ahead of Citibank's. As was true for AT&T, Bank of America, and United Airlines, though, strategy and technology were not enough. Sears and American Express are both firms whose vision is clear and commitments aggressive and effective, but in each case sub-stantial internal doubt and competition have gotten in the way of turning strategy into results.

One reason for asking the question "What makes our firm spe-cial?" is that focusing on the strengths of a culture builds the link between business idea and organization base. In the same way, flush-ing out agreement at middle levels of the firm looks at change as an opportunity rather than a problem.

Giving Permission to Take Risks

In many companies, including Citibank, American Hospital Supply Corporation, American Airlines, and others where top management played central roles in creating the momentum for change, innovation still relied on initiatives from lower down in the firm (and in many cases on inspired luck). In many other organizations, though, there are plenty of good ideas about telecommunications but a paralysis about getting moving.

The difference relates partly to the explicit or implicit signal senior managers send to their people that they will not be penalized for tak-ing risks. Telecommunications involves many uncertanties and risks plus long lead times. The middle level manager who wants to launch a new electronic product, the telecommunications manager who argues that the firm must build substantial extra capacity now to be positioned to handle unknown growth in services, or the sales executive who takes the gamble of giving the field sales force personal computers linked to special databases—all are stepping outside the traditional bounds of responsibility and combining business and technical decisions. The decisions may not work out.

Do they hold back because, as one manager in an airline commented, "You have to attract attention to innovate in an area of telecommunications that you are not fully knowledgeable about. In our company, there is a general feeling that management has a long memory. If you stood up for a program that didn't work, that's on your record—or at least in management's memory."

The consequence of this feeling is seen in at least half the Fortune 100. Business managers expect someone else to take the lead: top management, corporate information systems, or business planners. This leads to costly blockages:

- *Too much emphasis on central planning.* Where do ideas for new products come from? The central telecommunications unit often ends up taking on a business planning role that should be handled by the people close to the customer.

- *Short-term and reactive thinking in a field with long lead times.*

- *Loss of creativity.* In effect, the firm automates the status quo, and middle-level business managers will buy into projects such as office technology that have a senior champion but avoid being the champion themselves.

The most striking, if not always successful, firm in making sure that the middle of the organization played a proactive role in innovation for the electronic marketplace was Walter Wriston's Citibank, which prided itself on its philosophy of "Let a thousand flowers bloom." Perhaps there were too many simply dumb ideas, botched projects, and badly designed systems as a result, but there was also a tremendous number of small, medium, and large successes because Citibankers could feel that they would be positively remembered by management for taking risks.

Thomas Theobald, Citibank's head of international banking, describes very accurately the process and the impact of the environment for innovation he helped create in reference to the complex electronic banking system that became the springboard for "Citibanking":

> A very strong-minded, uncooperative type of operations manager, acting basically on his own, began to develop some of these systems. He was difficult . . . he would stand up at a meeting and tell the chief exchange trader that the trader did not know anything . . . but he, operations manager, would tell him in due course when he delivered the new system. If that kind of environment did not have some tolerance for conflict and very unorthodox organizational initiatives, the operations manager would have been fired. But his boss supported him and allowed him to spend a lot of money.

Theobald allowed, even encouraged, several competing projects and eventually put his weight behind the maverick's system.

> But what happened to the other two ideas? The people who were sponsoring and spending their money on them both got promoted, because they had taken the risk. . . . Everyone knew it was no disgrace to have something innovative rejected.

In several of Citibank's major competitors, there was a split between adventurous thinking at the top and caution below. For instance, Manufacturers Hanover's Geonet was, and is, far better designed and cost-effective than Citibank's worldwide communications network. Senior managers took some significant risks and the telecommunications team had a bold technical vision. Most of the organization, though, played a passive role in exploiting Geonet. As a result, the business traffic did not grow. Chase similarly had some excellent highway plans and planners, but little momentum at middle levels to come up with ideas for traffic.

Who comes up with initiatives for new electronic products? What incentives are there for managers to be champions in an area that is not part of their direct responsibility? How do they view the rewards and overt and covert penalties for taking the risk?

Innovation does not just happen. If innovation were easy, everybody would have done it. If the competitive gain were guaranteed, everyone else would have moved, too. If we can buy a package off the shelf, so can anyone. The sequence for integrated information technologies requires a strong push from the top so that people at all levels can contribute to the principal ingredients defined throughout this book: vision, policy, and architecture. Mobilizing begins when people share the vision, the culture buys into it, and people believe that they can and should make a contribution. It continues with education so that they are able to make an informed contribution.

In every single step, there has to be top management commitment and involvement. The issue is not one of time but of attention; in most instances senior managers have to provide direction, unblock constraints, send signals, and approve or supply resources. They can then get out of the way and let others get on with the lengthier process of planning and implementation.

They do not have to spend much time. They do have to pay attention. They also need to make sure that their business message is communicated and understood. That needs education.

STRATEGIC EDUCATION

Almost a prerequisite for pushing commitment down through the organization and mobilizing the interest and ability of people at all levels is strategic education. This is not the same as training. Too often, training is provided after a project has been implemented and aims at speeding up adoption of a system. For example, one-day courses on personal computers, seminars on how to sell cash management products, or self-instruction modules on using word processors follow installation of the relevant technology.

Education, however, is in no way the same as training. Training focuses on the skills needed to use the systems. Education is a mobilizing force that precedes development. It must have behavioral objectives—what do we want people to do, not what is the content of the program. The purpose should be to:

- *Change attitudes* and build awareness of the business opportunities from IIT
- *Share information* and improve communications
- *Build skills* for participation in planning
- *Stimulate action*

A strategic education program needs guiding principles. In order to lead change, education must be based on a clear idea of what the changes are intended to achieve in business and organizational terms. The education will be invaluable if it clears away fog and sets realistic expectations. To create a climate and context for action, education must show people personal benefits from change. In some cases, change is a matter of survival. In most firms, though, the picture is more blurred: "Yes, technology is important, but so are many other things. Why should I spend time on it?" The firm's managers and workers must learn to see change as a personal and corporate opportunity.

Education must make the abstractions of the technology concrete and meaningful. Relying on exemplars from the industry helps. Demonstrating management's commitments, communicating the business vision, and clarifying the policy decisions will gain cooperation at all levels of the management ladder. The presence of a senior manager at a seminar or workshop, even for a short time, demonstrates real commitment as no memo can. By providing a forum for people to express their concerns openly, education fosters true involvement and melts the resistance that often reflects skepticism or misunderstanding rather than overt opposition.

Because education is very expensive, taking busy managers away from their jobs for days and adding to their responsibilities when they return, it has to be justified on tough business criteria: Will this help us move faster? Will it lead to better systems and more effective implementation in terms of behavioral objectives, or is the money better spent elsewhere? Much of the expenditure on IIT training is wasted because it does not mobilize for action but just informs for interest or entertainment.

In the financial service industry, the leaders in the electronic marketplace spend about $1,000 per employee for each of two years once they recognize that they need to build organizational momentum behind the business and technical strategies. Sometimes the investment is part of a catch-up program. In particular, they find that their sales forces—account officers, dealers, and brokers—do not know how to sell electronic products and do not share the vision. Training them to use the products or be able to demonstrate them to clients is too little, too late. They overlooked the need to create awareness and understanding, change attitudes, share information, and create action *before* they launched the projects.

Financial services have been the bellwether for many aspects of telecommunications for business strategy. The level of expenditures on education and the cost of acting too late have messages for other sectors. In manufacturing, there is rarely sustained early education as part of the mobilization for office technology. There is plenty of training about how to use word processors, personal computers, electronic mail, and so on, but again, too little in most instances and too late.

The Education Development Process

Not only is education expensive but the development process can be complex. Training programs can often be bought off the shelf. There are plenty of first-rate firms, good teachers, and alternative mechanisms, ranging from seminars to video-based programs and computer-aided instruction.

When the aims of education are to mobilize for radical change, however, they need to be more focused on the organization, rather then the topic. A common problem for business managers is that they find it hard to link the messages about the technology to their own situation. If education builds awareness and unfreezes attitudes without at least pointing the way to action, it can be a source of frustration rather than a force for building momentum. Many of the better university programs for executives suffer from this limitation. The content is relevant, the teaching effective, the messages valid, but they do not connect to the manager's own world.

If education is propaganda, the gap between theme and action can be widened, not bridged. Vendors of IIT products, particularly IBM, have understood the importance of education in their own marketing. IBM's executive education programs have helped it get an access to top managers that no other vendor has been able to copy. The leading software providers frequently sponsor one- or two-day programs that draw on top academics, consultants, and respected senior IIT managers.

Many of these vendor-supplied education programs are valuable in building awareness and most of them avoid propaganda. But they naturally present a particular view of IIT and management options. They can be unsettling, too, since they sometimes send a signal to business executives that there is only one right way to do. They highlight risks and options but do not help executives choose between them.

It is far too easy to pick out what the IIT field is doing wrong. Every firm is making mistakes and will continue to do so. It is part of the uncertainty, immaturity, and risk inherent in mobilizing for change—and necessary for the return. If it were easy to do, every firm would be doing it. The leading influencers of management opinion have a responsibility to help, not just knock.

Public seminars, vendor-supplied workshops, university programs, industry conferences are all good sources of ideas and information, but they complement not substitute for a sustained strategy for education across the firm.

The Organizational Communities

The first step in mobilizing the organization for change is to identify the main communities whose understanding, attitudes, and actions will affect the success of implementation. "Communities" are groups of shared perspectives, values, and responsibilites. For example, senior management is mainly interested in finding out about the long-term business opportunities from IIT, the policy options, and the critical areas for decision-making. Middle managers are far less concerned with policy issues and need more direct guidance on what they can expect to happen in the next few years, and on what the impact will be on their own costs and operations. Senior managers are rarely interested in demonstrations of office technology tools or the new electronic products the firm plans to launch.

Middle managers want such demonstrations. They need to understand just what the tools are and get a sense of where and how they can ensure that there are no problems of disruption in their introduction. Senior managers often view the issue of how to allocate costs as a minor one. For middle managers, it is central.

The education process has to recognize the different situations and needs of the different communities. The specific communities vary depending on the organization, but for most firms they correspond to the list below.

- *Senior managers*: Telecommunications and business strategy, managing technology, strategic issues and options, priority next steps, the economics of technology capital

- *Middle managers*: Introducing technology into the workplace, the firm's strategic agenda, implications for management, getting business value from existing technology

- *Corporate planners*: Competitive trends, implications for financial planning, forecasting business needs

- *Technology architects and planners*: Competitive uses of integrated information technology in our industry, strategic technology trends, standards

- *Telecommunications and information systems staff*: The new business and organizational environment for the IIT organization, managing organizational change, strategic trends in technology and applications

- *User liaison staff*: Getting ready for X (office technology, the new EFT/POS project, branch rationalization, etc.), making systems work: roles and stages in effective systems development

- *User staff*—or rather, the IIT clients and colleagues: Getting ready for X: what it means for you, your own role, timetables, and training

In general, the starting point for a strategic education program is a top management workshop that gets senior executives to look at the issues of IIT as a whole, to recognize their own role in directing change and making "commitment" meaningful, and to raise their awareness that delegation is not the strategy any longer. That is only the start. One of the difficulties in mobilizing for radical change is that two forces have to work in parallel:

Push Commitment Down from the Top and Push Experiment Up

If the strategy starts at the top, moving awareness of commitment down moves at best one level a year. The strategy loses its momentum. The vision becomes blurred. If the education process moves senior management to move, it has to be complemented by mobilizing the middle levels at the same time.

If education creates interest and action at lower levels without corresponding policy decisions at the top, the result is at best frustration. Balancing top-down commitment and bottom-up experimentation is one of the keys to stimulating radical change.

Designing the Specific Education Programs

The specific topics are less important than the process of development. The problem is how to connect the general issues of business and technology to the specific context of the firm. This requires two resources: organizational "snapshots" and a special teacher. The shapshots are data drawn from the firm—quotations and examples. Even when the workshop comes off the shelf, whether a consulting firm's or university professor's, someone has to interview a representative cross-section of the main communities and find out what their concerns are, how clearly they understand and believe in the firm's vision and commitments to IIT, how much they themselves understand about technology or want to understand, and their own ideas about what a strategic education program should aim at accomplishing.

The results are often surprising to top management. They say to every senior manager, "You may know how important IIT is to your firm and may be fully committed, but your people do not share your view. They still feel there is a big culture gap between your technology organization and the rest of the firm. Their attitude is more often one of apathy, skepticism, and concern than of informed enthusiasm."

The special teacher must combine four areas of skill, being

1. An *instructor* with good presentation skills, the ability to lead discussions, and knowledge of the subject
2. A *researcher* with knowledge of what is happening in the IIT field
3. A *consultant* with experience in the field, credible recommendations, knowledge of the industry, and experience in working at this level of large organizations
4. A *change agent* with understanding of the politics of change, and the ability to unfreeze people's thinking

Tact with candor is needed in order to help organizations help themselves. There is no large firm that is doing everything right in IIT, not even the exemplars. Change is frustrating, difficult, and painful. Teachers who are too tactful do not force participants to face up to the dilemmas of change. Those who bludgeon them with criticism do not help them build the confidence that the technology can be tamed, and that the need for radical change is an opportunity rather than a problem.

11 HOW TO MOVE FAST

CATALYST: A FRAMEWORK FOR CREATING THE TIME RESOURCE

This final chapter addresses the obvious question that any busy senior manager must ask if he or she accepts the main line of argument of this book: What do I do now? The book has shown that time, not technology, is the strategic management issue. The lead times for building the communications infrastructure and communications-dependent services are long. The cost of following too late adds urgency to the management agenda. Being there when demand takes off means making choices now to be in place two to seven years from now.

Investing in telecommunications and integrated information technologies requires resources of capital, technology, management, and time. Firms cannot buy time off the shelf. When the other components of the investment base are in place but the company is not competing in time, the strategy is largely irrelevant.

This last chapter presents ways to create the time resource. It is based on a methodology called Catalyst[sm] that I have developed as I have worked as a researcher, teacher, and advisor in helping large firms take charge of change and in learning from them. I have applied the components of Catalyst in a number of organizations, the testing ground for validating the realism of the ideas and practicality of the recommendations expressed throughout this book and also the framework for updating and adding to them. Working with my colleagues at the International Center for Information Technologies, I have formalized what was previously a coherent and flexible, but largely intuitive, approach to taking charge of change, and Catalyst is now a formal diagnostic tool that highlights the priorities for action and the best vehicles for gaining time. Figures 11–1 and 11–2 contrast the standard view of telecommunications planning with the Catalyst perspective.

You can use Catalyst informally. It is mainly a checklist of activities that can help gain your organization time. Catalyst involves three steps in analysis:

1. Assess where you stand in relation to the seven factors that need to be combined to ensure that a sound technical or business strategy for exploiting telecommunications is not blocked by a weak and ineffectual organizational or financial one, or vice versa.
2. Identify how to get ahead of the game and think ahead, in order to define "springboard initiatives."
3. Identify how to accelerate the pace of change in effective definition, delivery, and use of the springboards.

The first step is a very short reality-test, an assessment of where your organization as a whole or your own strategic business unit stands in relation to the seven factors that distinguish the winners from the losers in competing in the electronic marketplace. This step orients you by helping you to make sure that the interrelated managerial, technical, organizational, and financial issues are not treated separately.

In addition, it clarifies whether the focus in the next steps in the Catalyst analysis should be on identifying "springboard initiatives" and expanding your firm's scanning of its opportunities and/or competitive necessities, or on taking action to speed up their delivery and use. The lower your rating of senior management awareness, the more likely it is that you need to focus on choosing your springboard initiatives. The higher the rating, the more likely it is that you need to focus on the "how to get them in place fast."

"Springboard initiatives" are the major moves to be made in applying telecommunications in your organization. These are the moves in which time is most crucial. Catalyst is then used to assess the most effective way to gain that year, by asking "Is this specific activity likely to help speed up our progress and remove a bottleneck to gaining time?"

I chose the word "springboard" rather than "strategic" as a reminder that many investments in the telecommunications, computing, and information infrastructures are the enabling system for a

Figure 11–1.   The Standard View of Information Technology Planning.

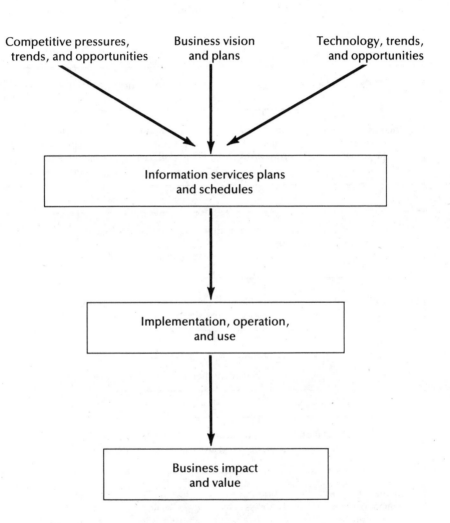

Figure 11–2. The Catalyst Perspective.

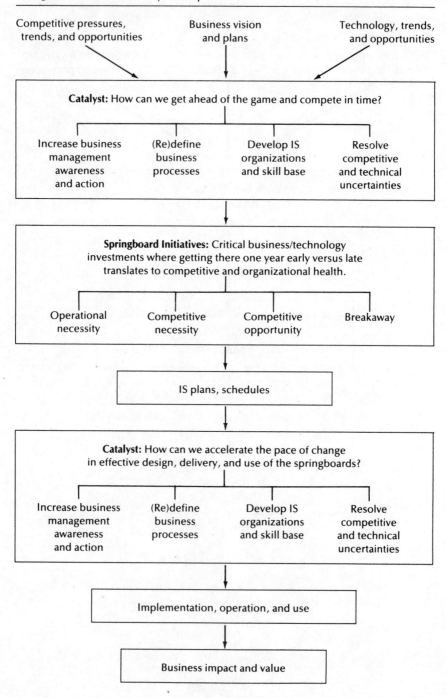

business benefit, the springboard for aggressive action. They may in themselves not seem at all strategic; that is why they are so easily overlooked in the formal planning process. (As was pointed out earlier in this book, this is especially true when they relate to using telecommunications to differentiate old services rather than create new ones.) The underlying question here is: What do we have to think about putting in place now so that we can build on it as part of later innovation?

You can assess the opportunities to think ahead and move ahead under four headings:

1. Increase business management awareness and action.
2. (Re)define business processes.
3. Develop the information services organization and skill base.
4. Resolve business and technical uncertainties.

STEP 1: RATING YOUR ORGANIZATION ON THE FACTORS THAT DISTINGUISH THE WINNERS FROM LOSERS IN THE ELECTRONIC MARKETPLACE

The first step in taking charge of change is to recognize where your firm or unit stands in relation to the seven factors that I have found consistently differentiate the winners from the losers in competing in the electronic marketplace. They were mentioned briefly in the introductory chapter. Rate your firm—or division or business subunit, depending on your focus and interest—for each of the factors on a scale of 1 to 5.

Rating

1 = We are weak here, and this weakness is a significant disadvantage to us. If we do not fix it, our long-term competitive positioning will be seriously affected.
2 = We are weak here, and it is important to fix it if we are to make progress.
3 = We are in good shape here, and this does not require special attention or major interventions to improve it.
4 = This is an area where we are strong and which we can build on and exploit in our strategy and implementation.
5 = We stand out here as special and should be looking for ways to exploit it to the fullest.

Factor

1. Senior management awareness and willingness to turn awareness into action
2. Understanding of customers' actual motivations and behaviors
3. A defined architecture and architect
4. Middle management buy-in
5. A seven-year thinking horizon and follow-through
6. An integrated information services organization
7. The management climate to support courage

Each of these factors is summarized below, together with vehicles for change. The value of focusing, even quickly and subjectively, on how you rate your business unit (and yourself and your colleagues) is to avoid overcompartmentalization. The rating will help you to see important interdependencies or blockages. The most common blockage is shortsightedness. There are many other oversights because of compartmentalized thinking. Very few plans that focus on competitive opportunities, for instance, at the same time pay attention to the biggest single problem in many firms: getting middle management to "buy in." Middle managers are seeing and will continue to see a 20 percent reduction in their ranks. Their budgets are being tightly controlled or cut, but the work loads go up and not down. They are the ones funding the typical 10–25 percent increase in computers and computing investments; information technology is the only major area that is growing faster than the rate of business growth.

Middle managers are also often seeing the end of the business axiom that underlies their long-term career history. This is the assumption that experience is valuable. Telecommunications often invalidates experience. The comment of a senior Citibanker around 1981 seems prescient: "The credit people believe that credit is hard to learn and that therefore experience matters. They don't yet see that electronic banking means they will have to learn a lot about technology and about selling services they don't understand. Experience won't help them much but they think it will. That's why they won't bother to learn till too late."

Telecommunications professionals are experiencing this invalidation of experience, too. Five years ago, to have a resume which showed 20 years solid background in voice communications in the old era of regulation and Bell System monopoly meant a lot. It is now seen as a disadvantage, rather than an asset, as firms look for people with up-to-date knowledge of modern digital technology and data communications in the entirely different "postdivestiture" circus of today.

Without middle management's understanding of, belief in, and benefit from the changes telecommunications creates to their work units and their own jobs, organizational inertia and loss of morale make it hard to turn even the best business and technical plans into effective outcomes.

Middle management buy-in is of growing importance in terms of managerial responsibility and challenge and it highlights the vital need not to concentrate on getting the business strategy clear plus the technical strategy plus the funding strategy while ignoring the organizational strategy. The same is true for any other combination of the seven factors.

Traditional IS planning too easily builds false compartments. Catalyst breaks them down.

Factor 1: Senior Management Awareness and Action

Senior management's awareness opens up the possibility and likelihood of action. Senior management action adds momentum and credibility to others' action. This has been the organizing theme of this book and the major message in just about every leading consultant's, academic's, and practitioner's recommendations.
Management needs to

1. *Build the business vision*: a photograph of the future, a focused business message to drive the entire planning and decision process for telecommunications and to help determine where the firm should lead and where we should follow, and where we should compete and where cooperate and look for strategic alliances inside or outside the firm's traditional industry.
2. *Make explicit policy decisions* on applying the technology to core business drivers and on appropriate business justification.
3. *Show "commitment"*, in terms of action, not memos, and behavior, not goodwill.

Some general vehicles for change are made more specific in the next step of the Catalyst analysis. They include:

1. *React to crises*—lose position to old and new competitors. This concentrates the managerial mind wonderfully but rather expensively.
2. *Look for exemplars*—track the best of current practice, scan and communicate competitive trends in your "industry" and near-industry.
3. *Educate and communicate.*
4. *Build mutual understanding* between your business leaders and information services managers and establish forums for working together at senior levels, with the information services manager acting as a gatekeeper and translator to senior business management.

How do you rate your organization on this factor? 1   2   3   4   5

Factor 2: Understanding Customers' Actual Motivations and Behavior

The need:

1. *Focus on customer research* rather than market research.
2. *Encourage field-centered initiation* of product and service development.
3. *Base innovation on customer benefits.* Ensure that any communications-dependent business innovation be based on self-explanatory and self-justifying customer benefits.
4. *Build and maintain close and direct contacts with customers* by the teams that propose innovations.

Vehicles for change:

1. *Build customer relations databases* that identify each customer uniquely, provide demographic data, profile product use and history, and clarify the profitability of the relationship.
2. *Focus on the core drivers of the business* and primary convenience service. Make it easy for the customer to deal with the firm in basic transactions.
3. *Scan exemplars* for messages about customer responses to moves in the electronic marketplace; recognize that technology is no respecter of traditional industry barriers.

Your rating? 1   2   3   4   5

Factor 3: A Defined Architecture and an Architect

The need:

1. *Ensure a corporate view* of the communications infrastructure.
2. *Define a core set of standards*—technical and business-based— for the interface of a multiservice workstation to the information utility.
3. *Define the priority data resources* to be shared and/or stored in a standardized form.
4. *Create a phased blueprint* for moving toward integration.
5. *Assure authority.* Make sure that the manager and group accountable for the integrity of the architecture have the matching authority to ensure that the standards have teeth.

The vehicles for change:

1. *Publish the focused directional business message* and statement of vision to provide criteria for defining the arhitecture.
2. *State a clear top management policy on who decides what.*
   a. Centralize coordination of the infrastructure.
   b. Decentralize development and use.
3. *Publish the selected strategic standards,* and ensure that business units recognize they are not just guidelines but standards with teeth.
4. *Provide a one-page summary of the architecture*: the architecture is the strategy.

Your rating? 1   2   3   4   5

Factor 4: Middle Management Buy-in

The need:

1. *Establish belief and benefit,* through choice of innovation, education, incentives, and demonstrated, credible management commitment.
2. *Permit and reward risk-taking.*
3. *Foster cooperation.* Make sure the IS and business units, and field and head office move toward a "we together" attitude and not "them versus us."
4. *Support the new customer-focused sales relationship.* Provide resources and expertise to create a new selling culture to support

relationship management, rather than product-by-product selling.

Vehicles for change:

1. *Provide sustained and public senior management action:*
   a. Push commitment down one level each year.
   b. Pull experimentation up.
   c. Use your behavior, not memos, to show commitment.
   d. Ensure natural, direct, and continued communication across the culture boundaries.
2. *Create an organizational plan for change*, as well as a business and technical one:
   a. Provide early education not late training.
   b. Review incentives and clarify career growth opportunities.
   c. Build a cadre of hybrids.
   d. Insist on distributed management authority and responsibility for distributed information technology development and use.
   e. Look for opportunities within the architecture to deliver high-payoff, fast development applications that provide benefits to middle management.

Your rating? 1   2   3   4   5

Factor 5: A Seven-year Thinking Horizon and Follow-through

The need:

1. *Manage the information technology investment as a capital asset* not an expense.
2. *Provide corporate funding for the long-term infrastructures* and enabling systems.
3. *Focus the business planning on providing the vision* so that you can be there when demand takes off.
4. *Develop a product stream to exploit the architecture.*

Vehicles for change:

1. *Ensure vision, policy, and architecture.*
   a. The vision is defined and communicated.
   b. The policy is clear.
   c. The architecture is defined.

2. *Establish a Gang of X*—a small group combining some seniority, some innovation, some technical knowledge, and plenty of organizational credibility to provide direction and strategic review.
3. *Treat technology investments as business capital investments.* Establish mechanisms at the corporate level and within business units to ensure a view of the technology investment as a business capital portfolio.
   a. Require systematic business justification of radical, innovative, incremental, and operational moves in appropriate business terms.
   b. Fund the infrastructure.
4. *Eliminate short-term, cost-based allocations* and move toward a quasi-profit center pricing strategy for Information Services. Create a regulated free market for information technology.
5. *Formalize business/technology competitive scanning across business units.*
6. *Set up a business-centered information systems R&D capability,* for funding pilot projects, working with customers, and scanning key technologies and uses.

Your rating? 1   2   3   4   5

Factor 6: An Integrated Information Services Organization

The need:

1. *Ensure integrated thinking* in an integrated delivery organization to manage integrated technologies.
2. *Locate, encourage, and reward hybrids* and build liaison roles linking IS and its clients, colleagues, and partners.
3. *Develop a new generation of technical specialists* who can contribute to the integration of the technologies rather than just be experts on their own component of the integrated information utility.
4. *Build a cadre of real managers* instead of promoted technicians who learn by vicarious trial-and-error failure.
5. *Define new methods for getting IS work done*:
   a. Business-centered planning.
   b. Business-led definition, design, and delivery of information products and IS support.

   c. User-led projects, rather than just IS-led management of the technical work.
   d. Fast delivery development tools and investments in ways of improving all aspects of software productivity.

Vehicles for change:

1. *Management energy*: Skunk works, tiger teams.
2. *IS Organization*:
   a. Bring telecommunications and information systems together.
   b. Mission statement and "news".
3. *Career redefinition*:
   a. New roles.
   b. Self-development.
   c. Crossfertilization and lateral development.
   d. Education plus training.

Your rating? 1 2 3 4 5

Factor 7: A Management Climate to Support Courage

The need:

1. *Avoid automating the status quo.*
2. *Encourage and reward suitable risk-taking.* (Remember: If innovation were easy, everyone would be doing it.)
3. *Show your leadership.*

Vehicles for change:

1. *Management action versus party line.*
2. *Rewards and recognition.*

Your rating? 1 2 3 4 5

How do you rate overall? Completing this short self-inventory gives you a fairly clear overall picture of where you stand and the major factors your firm or unit can exploit versus those it needs to improve and those that are significant blockages to progress. Every one of them can be influenced and in many cases transformed by senior business management action. None of them can be changed from weak to strong without that action.

STEP 2: MAPPING SPRINGBOARD INITIATIVES

You can use the next steps in applying the Catalyst framework for analysis at several levels of detail, from a one-day brainstorming by a management team to a very detailed study. Catalyst should not, though, be used for a lengthy study. The need is to turn vision into policy into sustained action quickly. Of course, if the analysis shows that the best vehicle for accelerating the pace of awareness and action is, say, a short education program for business unit managers, plus a revision of the financial justification process plus a competitive scan of potential third-party competition, each of those activities takes time and resources to plan and deliver. Catalyst aims at getting them identified early and acted on quickly.

The second step in the Catalyst analysis addresses the issue of making sure you are not just automating the status quo. What are your springboard initiatives? What should they be? These are investments and projects that everyone with responsibility for the organization's economic and organizational health recognizes—or should recognize—as having important consequences if they are put in place earlier versus later. They may relate to competitive opportunity or competitive necessity. They may be radical, innovative, incremental, or operational moves. They are the responses you can and should make to

1. Competitive pressures, trends, and opportunities
2. Business vision and plans
3. Technology trends and opportunities

If you rated your organization as less than a 4 in terms of senior management awareness and action or your management would be unfamiliar with more than just a few of the examples given in this book of how firms have *already* used telecommunications for business advantage, it is virtually certain that your firm's current business and technical strategies are more reactive than they should be in terms of telecommunications and that you need to ask, How can we get ahead of the game and think more imaginatively about the business options? Are we like the well-run oil company where senior management does not recognize the vital need for a retail point-of-sale and credit card strategy? Or the insurance firm whose executive committee recently cut back funds for agency automation? Or the international bank without an international network? Are we thinking ahead broadly and clearly enough? Do we need to be? The second step in Catalyst looks at mechanisms for answering these two questions. Figure 11–3 shows the checklist for this. It shows the four main targets of opportunity for accelerating progress:

1. *Increase business management awareness*, mainly through education (*not* training).
2. *Define or redefine management processes.* An obvious example here is to change the process of business justification and cost allocation so as to remove barriers to treating the telecommunications investment as business capital not overhead expense.
3. *Develop the Information Services organization and skill base.*
4. *Resolve competitive and technical uncertainties*, mainly through competitive scanning and research.

For each of the four categories and for the specific activities, you need to assess whether investing attention, time, and resources here is:

1. *Not a priority* in terms of competing in time. We can afford to continue on our current path since if we increase our efforts here it is unlikely to have much impact on where we end up 3 to 5 years from now.
2. *Desirable* to invest here; doing so will help accelerate the pace and effectiveness of change.
3. *Urgent and vital* if we are to accelerate the pace and effectiveness of change.

The checklist in Figure 11–3 is not exhaustive. It may well be, for instance, that you can identify other activities that merit a 3 as "urgent and vital." One firm, for instance, decided that the best way to accelerate progress was simply to acquire a software firm and at the same time to fire its IS manager. That took the brakes off imaginative thinking.

What you are almost certain to find if you rated your organization less than a 3 on senior management awareness, understanding of customer motivations, and having an architecture and architect is that you do not really know what your springboard initiatives are, could be, and should be. You may then view it as essential for someone (probably but not necessarily an outsider) to brief top management, to scan and redefine what you mean by "our industry" and "the competition," and to put new efforts into formally building and communicating the business vision. Perhaps you may feel instead that redefining the base for business justification of radical moves and investments in the telecommunications highways is the key to removing bottlenecks to thinking ahead. There is no fixed agenda or packaged set of programs you can apply.

You must obviously, though, have a base for deciding which actions are both important and urgent. This means you need to identify how you can help make sure your organization defines appropriate springboard initiatives. These are not just "important" but are ones where timing matters. They are also not necessarily "strategic." Many needed initiatives get overlooked because they do not grab attention at a strategic business level. They later turn out to have been essential to build a springboard. Many of them relate to operational systems and to basic infrastructures that are taken for granted but need renewal.

For example, for many banks the most important investment in information systems is not new development but redevelopment of existing ones. One bank that has been the leader in its national marketplace in retail electronic banking regards redesign of its demand deposit accounting system and Visa processing as by far its most important projects. These were designed 20 years ago for use in a much simpler business context, to handle far lower volumes, in a simpler technical environment—with no telecommunications. Reinvesting in these old systems is an operational necessity.

The same bank sees point-of-sale as a springboard initiative, for reasons of competitive necessity, and investment in a new internal network as a competitive opportunity. It sees no initiatives that are likely to give it a chance to make a breakaway. Your organization will have a similar mix of initiatives along this spectrum:

1. *Operational necessity.* Like it or not, we just have to do this to be efficient and responsive.
2. *Competitive necessity.* Like it or not, we have to do this, too, or we will fall behind competitors in terms of customer service, product differentiation, and so on.
3. *Competitive opportunity.* If we get moving on this now, we will be positioned to push others into competitive disadvantage and exploit occupancy and lead time.
4. *Breakaway.* Here is our chance for a preemptive strike or for substantially changing the dynamics of competition in our business sector(s).

Figure 11–3. Catalyst Checklist.

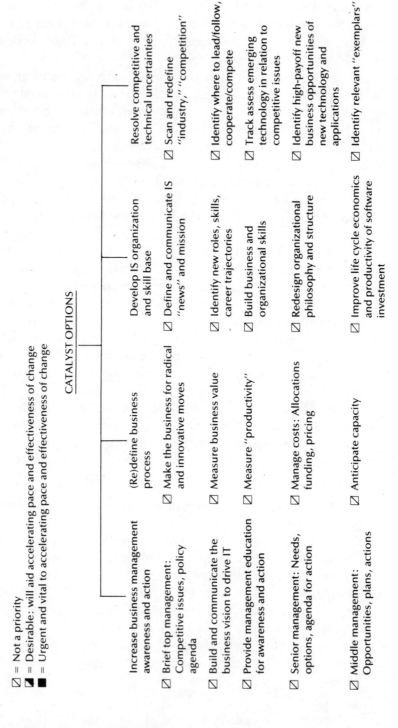

☐ = Not a priority

◨ = Desirable: will aid accelerating pace and effectiveness of change

■ = Urgent and vital to accelerating pace and effectiveness of change

CATALYST OPTIONS

Increase business management awareness and action	(Re)define business process	Develop IS organization and skill base	Resolve competitive and technical uncertainties
☐ Brief top management: Competitive issues, policy agenda	☐ Make the business for radical and innovative moves	☐ Define and communicate IS "news" and mission	☐ Scan and redefine "industry," "competition"
☐ Build and communicate the business vision to drive IT	☐ Measure business value	☐ Identify new roles, skills, career trajectories	☐ Identify where to lead/follow, cooperate/compete
☐ Provide management education for awareness and action	☐ Measure "productivity"	☐ Build business and organizational skills	☐ Track assess emerging technology in relation to competitive issues
☐ Senior management: Needs, options, agenda for action	☐ Manage costs: Allocations funding, pricing	☐ Redesign organizational philosophy and structure	☐ Identify high-payoff new business opportunities of new technology and applications
☐ Middle management: Opportunities, plans, actions	☐ Anticipate capacity	☐ Improve life cycle economics and productivity of software investment	☐ Identify relevant "exemplars"

☑ Middle management: Getting business value from existing systems

☑ All levels: Understanding telecommunications

☑ All levels: A vocabulary for IT planning and action

☑ Alert senior management to new competitive threats, opportunities

Build plans for managing organizational change

  ☑ Middle management jobs

  ☑ Impacts of work, careers

  ☑ Education and career development

☑ Other (list) _____

☑ _____

☑ Make "involvement" work

☑ Steer committee agenda and actions

☑ Imbed IT planning in business unit

☑ Define and coordinate international strategies and operations

☑ Other _____

☑ _____

☑ Educate TC, DP professionals about new technologies and integration of IT components

☑ Articulate business capacity needs

☑ Other _____

☑ _____

☑ Fix baseline technology assumptions for 1992, 1997

☑ Define integration plan

☑ Other _____

☑ _____

STEP 3: HOW CAN WE GET OUR SPRINGBOARD INITIATIVES IN PLACE ONE YEAR EARLIER?

Some firms already have a good idea of what they need in each of these areas, especially if senior management awareness is high and has been turned into visible commitment. For them, the third step in the Catalyst analysis is most relevant. This uses the same checklist as in step 2 but focuses on the question "How can we accelerate the pace of change in effective design, delivery, and use of the springboards?" Is the best way to increase business management awareness and action, to define or redefine business processes, to develop the IS organization and skill base, or to resolve competitive and technical uncertainties?

Here, as in step 2, you need to assess each specific activity on the checklist on a 1–3 scale and also to identify other ones that are a 3. When you match the assessments made in step 2 with those in step 3, where an activity was rated a 3 in both phases, it is without any doubt a priority for action. It is seen as both urgent and vital in terms of identifying your springboard initiatives and getting them in place.

The Catalyst analysis can be done quickly. The follow-on implementation of the activities needs time and resources. They will generally include education and research. Your organization probably already invests in these. The information services field is probably the most research-receptive of any business function. However much IS managers complain about impractical academics and the cost of consultants, they rely on them for help in resolving technical and business uncertainties, in keeping them up to date, and in educating them and their staff. Management education is the fastest growing part of the IS budget in leading firms.

What Catalyst helps do is provide a clear basis for investing in education and research and making trade-offs between that and other aspects of planning and organizational development. It also helps clarify the goals for education. These should not relate to topics but to action and not be defined in terms of what the program is about but what it is meant to lead to. Quite literally, education should buy time—not buying it in the traditional meaning of the phrase, which is to get the opportunity to delay decisions or events, but buy it through getting ahead of the change curve, to be thinking ahead and moving ahead, and to be building senior management awareness and action in a positive way, instead of having them be more painfully created by competitors' successes.

The same is true for research and for redefining business processes. The objective should be to buy time. I chose the term Catalyst for the simple process briefly described here very consciously. A catalyst makes a chemical process happen but is not changed by it. It is an enabler. That is what most good organizations need. They have the management talent, energy, intelligence, and drive to take charge of change. They need catalytic forces to help them mobilize. I hope this framework can help your organization do this.

THE NEXT DECADE OF OPPORTUNITIES

This has been a very conservative book emphasizing what has been done within existing telecommunications technology. I have avoided technobabble and future speculation. Everything I have discussed is practical today and someone has already made it happen. But I want to end with a few less earth-bound recommendations. Neither the technology of telecommunications nor its business uses nor business itself remain static. If we go back to 1978, how could any of us have predicted 1988? Globalization of financial markets, the break up of AT&T but not IBM, the progress and problems of deregulation, the rise of Asian economies that threaten even Japan, the impact of the electronic marketplace, personal computers, the transformation of the telecommunications industry, etc., etc.

Nineteen hundred and ninety-eight is at least as different. Surely, none of us expect a slowing down in the pace of business and technical change. 1990 is almost here. We have only a few years to be thinking about 1995 before we have to make some business acts of faith concerning telecommunications and business advantage. Here is my own assessment of the next decade of opportunities, based on straws in the wind or the lucky advantage I have had in working with some of the very best thinkers and doers in business and academia, whose thinking aloud helps alert me to the shape of things to come:

Telecommunications. Telecommunications technology is just a means to a business end. New means open up new ends. There are several accelerating trends in the technology which will be used by the leaders in the electronic marketplace to show the followers what American Hospital Supply, Citibank, and American Airlines showed in the Middle Ages of telecommunications. Very small aperture terminals (VSAT) linked by satellite will be to telecommunications what the personal computer has been to computing. ISDN will take decades to put in place but isdn is already here. Electronic document management will be the luxury of 1988 and the necessity of 1990.

Multinationals.   The ability of a multinational corporation to manage complexity and volatility will depend almost entirely on its telecommunications architecture. Telecommunications will speed up the emergence of what I call the "federated organization," which is the emerging trend in industry after industry, and which is made possible only through telecommunications. In the 1990s, large firms in volatile environments will recognize that their ability to add, relocate, redefine, and divest organizational units will make the corporate backbone network central to flexibility and fast response. The telecommunications-based organizational structure will in itself become a strategic business asset.

Organization.   The focus of attention and payoff in investing in integrated information technologies will shift from economic health— gaining a competitive edge or avoiding a competitive disadvantage— to organizational health. In particular, telecommunications will be a major force in ensuring organizational simplicity and dealing with many common pathologies that impede many large firms' effectiveness. These include tensions between the field and head office, depersonalization of management, fragmented understanding, inefficient teamwork and project work, and subservience to documents.

Cities.   Telecommunications will have an immense impact on how cities compete. To be a major player in the international environment, a city needs to be an electronic banking and trading center, to attract multinationals, to provide open access to communications and not restrict business use, in order to increase or at least not remove business functions, and to be a major airport hub, when deregulation of the airline industry plus pressures to rationalize economics and operations forces moves toward the hubbing strategies that followed deregulation in the United States.

THE OPPORTUNITY OF EMERGING TECHNOLOGIES

The basic opportunity telecommunications has offered throughout its evolution has been to eliminate barriers of geography and time on service and coordination. This was true for the telephone. More recently data communications, by linking workstations to computers, or workstations to workstations, has broken down more barriers. The new generation of telecommunications technology extends this even more and thus opens up even more business opportunities, *given business imagination*. Three new capabilities it provides are *portability*, through VSAT, small earth stations that can turn a personal computer into a full-scale extension of the firm's information utility with no fixed cabling, *high-capacity transmission* links at low cost, via fiber optics and to a lesser degree satellites, and *multimedia information movement*, via electronic data interchange and document interchange software.

Each of these is already breaking down additional barriers of geography and time. Together with developments in computer and workstation facilities, they greatly expand the scale, scope, and effectiveness of a firm's information utility. They are all well out of the R&D stage but not yet fully proven or in widespread everyday use.

But . . . when do you lead and when do you follow?

VSAT will be to telecommunications what personal computers were to computing. Very small aperture terminals are 4-foot diameter satellite earth stations. If the FCC allows VSATs to take over unused radio frequencies earmarked for direct broadcast TV (DBS) they can be shrunk to 1 foot in diameter. VSATs can be installed in about half a day in most areas (except where there are problems of reception) and cost about $500 a month to operate. The current cost of a VSAT receiver is about $8,000 with prices dropping already. With volume production and use, they will follow the cost curve of other communications and computing hardware, typically a drop of 20–35 percent a year.

A typical VSAT system can receive data sent from the broadcasting hub at a speed between roughly 250,000 and 500,000 bits per second, fast enough for your firm to broadcast changes in its price lists daily. The local VSAT workstation can send data back at speeds of anywhere between 10,000 and 64,000 bits per second (64,000 is the new basic unit of measurement for high-speed communications use). Any type of information—telephone calls, video data, facsimile, numbers, documents, graphics, or even videoconferencing and product demonstrations—can be efficiently transported to and from a workstation at that speed.

VSAT is a proven technology but has still not really taken off, despite its promises. The reason is essentially because telecommunications managers mainly extrapolate from the status quo and look at VSAT in terms of cost reduction and substitution for local communications. They are making the same misjudgment as did data processing professionals who saw personal computers as a smaller and more amateur version of a "real" computer. It took users to lead the PC revolution.

It will need the same for VSAT to take off. Think of it in terms of end-user communications just as PCs are part of end user computing. Here are some business possibilities:

The Relocatable Office or Branch.   If your firm is a bank, how would you like to be able to open 200 ATMs at the Super Bowl for six hours, or operate a marginal branch in a small shopping mall at weekends and Christmas? If you are a construction or engineering firm, would it help to be able to put a high-powered personal computer out in the field, literally, and be able to link it to your home office computer facilities?

New Contracts.   How quickly can you connect up the dealer you have signed up for your agency automation capability, the customer for your cash management service? Three months if you are lucky. With VSAT you could leave the system in place when you sign the contract.

Temporary Installations.   Where would you gain from being able to get a full-scale computing facility up and running in under a day, no matter how isolated the location, with complete access to central databases and communications? For emergency relief work, sales activities, special projects?

Communication with District Branches.   Could you exploit enhanced communications for getting presentations or holding meetings with small remote branches—in fact with all your locations, so that you can get your messages across directly and quickly instead of by memo, rumor, or occasional state visit?

There are many practical problems in installing and using VSAT. They are the standard ones of international regulations, vendor reliability, critical mass, choice of subtechnology (such as choosing C-band versus Ku-band satellite broadcast frequencies). But, to quote

Peter Goodstein and David Sulser, "To paraphrase an old saying, waiting for VSATs to arrive is like waiting for the tide to come in. You can stand on the beach and get your ankles wet, and you still can't see the tide coming in. Yet, like VSATs, you know it's coming" ("The Business Opportunity of Very Small Aperture Terminals," International Center for Information Technologies, 1987).

When is it time to move? Do we lead, do we follow? The fact that we cannot use the technology immediately is irrelevant. If we hit on an imaginative business opportunity opened up by VSAT, it will still take us two to five years to put it fully in place. Two to five years from now, VSAT will be as widespread as PCs.

Similarly, the integrated services digital network (ISDN) will not be in place for a decade. This is the ultimate universal utility and the battleground for international control and the testing-ground for standards. Here again, though, do you have to wait? Any large firm can put in place what may be termed little ISDN. This is the corporate information utility that exploits existing standards to provide the full range of services within the company that ISDN will provide to all subscribers at some distant point in time and a broad range of services between companies. It allows firms now to exploit the advantages of high-speed fiber optics, wide-scale videoconferencing, electronic data interchange, and electronic document management.

This last area—electronic document management—is where many of the highest payoff opportunities for the next decade will be created. Computer professionals have a very narrow view of "information"; they think of it mainly as numbers in a "database" or messages. For most people explaining their work, information means documents. Many, perhaps even most, of any organization's mechanisms for coordination, management control, back-office customer service, inventory management, and so forth are built around key documents, which can almost take on a life of their own: forms, policies, way bills, purchase orders, and more.

One of the main reasons why computers and office technology have disappointed many firms, who expected but did not get measurable improvements in productivity, is that traditional applications have not had much impact on document movement and management. In a bank branch, for instance, shaving 5 cents off the cost of processing a check or making a credit card authorization has little if any impact on document-centered work, including updating, accessing, and using product manuals, paper-handling, and finding where information is stored.

The new technological base for telecommunications in the 1990s provides the capacity to move documents. Up to now, the speeds available were far too low for this to be practical. New computer-based technologies provide the capacity to store and access electronic documents and in fact every type of media quickly; CD-ROM (compact disk read-only memory) for instance, stores a 20-volume encyclopedia on a compact disk. You can browse information quickly and carry out the casual searches that would be impossible otherwise. "How many famous people in history have the same birthday as mine?" takes around 8 seconds to answer. CD-ROM has been available for several years. Like VSAT, it is an opportunity missing user leadership and imagination.

Standards are emerging quickly, too. IBM's Document Interchange Architecture provides a base technical standard for documents. Business-based standards are already stable or stabilizing fast; that is why electronic data interchange (EDI) has taken off in industry after industry, based on the X.12 standard.

Instead of waiting for ISDN, the leading firms are installing isdn. They are then positioned to evolve toward ISDN and to exploit the opportunity of electronic information-handling in its fullest definition. It seems foolish to wait.

TELECOMMUNICATIONS AND ORGANIZATIONAL SIMPLICITY

It is simply crazy for any multinational firm to wait. There is plenty of evidence already that telecommunications will be one of the main determinants of large, diversified, and geographically widespread organizations' ability to manage environmental complexity in the next decade. Telecommunications will also be a key element in helping business leaders personalize their leadership and in restoring organizational simplicity in an era where our organizations seem to be teaching a limit of complexity, but not a limit of size. Telecommunications is the base for organizational design and redesign in a time of discontinuity, not just rapid change. Here is the main opportunity of the next decade.

Large organizations in Europe and the United States are overcomplex and they become more so as they struggle to respond to the complexity of their environment, created mainly by three continued pressures:

1. *Globalization* and extended lines of communication, coordination, and operation across time zones and locations
2. *Hyperextension of activities* in terms of markets, services, products, customer demands, and anticipation of and response to competitive pressures
3. *Time stresses* created through geographic dispersion, shortening of planning, development, and delivery cycles, increased environmental volatility, and drastically reduced reaction times

This is the environment of the multinational corporation, now and for a long, long time. Firms have responded to the challenges of environmental complexity by adding resources for information management, in the widest sense of the term. This has mainly been through adding administrative overhead, formal control and reporting systems, layers of staff and management, and the substitution of impersonal paper for people in communication.

Computers have been a major force here, through management information systems and clerical automation. As a result, they have historically been associated with bureaucratization. The record of computing in large organizations through the early 1980s can hardly be defended as a force that has made them flexible and cut out unnecessary procedures.

We need to restore organizational simplicity and in particular deal with the common and pathological consequences of organizational complexity.

Tensions between the Field and Head Office.   Sales staff complain that they are not kept informed about products, have trouble locating the people who can answer their own or their customers' questions, or view the corporate office as a remote "They" who do not understand the fields needs.

Depersonalization of Management.   Senior executives do not meet their people except when they venture out on ceremonial visits; the CEO's business message and company news are sent via memo, not communicated directly.

Fragmented Understanding.   In a long chain of paperwork, stretched over time, people, and geography, no single individual may have a complete picture of the system but many may have vested interests and perceptions.

Teamwork and Project Work.   When work is distributed and lines of communication extended, coordination becomes costly and expensive; that is one reason why travel and documentation now add up to one-third of expenditures on a typical systems development project.

Subservience to Documents.   The obverse of the flexibility, face-to-face contact, and the single-step processing that organizational simplicity permits—a small firm in a single location, for example—is virtual ritualization of documents, claims, travel expense forms, etc., etc. can take over from the work process they are intended to support, controlling rather than being controlled by the people who need them transacted.

A Strategy for Organizational Simplification

Figure 11–4 shows a basic strategy for using information technology to break the link between environmental complexity and organizational complexity. The main principles apply to any component of the information services tool kit, from electronic mail to expert systems.

The strategy is to use information technology explicitly to:

- *Repersonalize leadership*
- *Increase direct, flexible contact between people*
- *Cut down on the need for information intermediaries*
- *Provide simple access to information*, simply organized
- *Focus on people's needs* for document-based information
- *Reduce document flows*, and barriers to tracking, locating, and controlling documented-based work flows
- *Get a firm drive from the top of the organization to cut layers of management*

The last point is not peripheral. Organizations do not delayer themselves, nor do well-entrenched administrative procedures relax under the logic of simplification. They are too often dominated by social inertia, which dampens effort to change them. Many firms have successfully used IT to increase organizational simplicity, but only when the leadership adds its weight to the push. The leaders' goal has obvi-

Figure 11–4.   Information Technology and Organizational Simplicity.

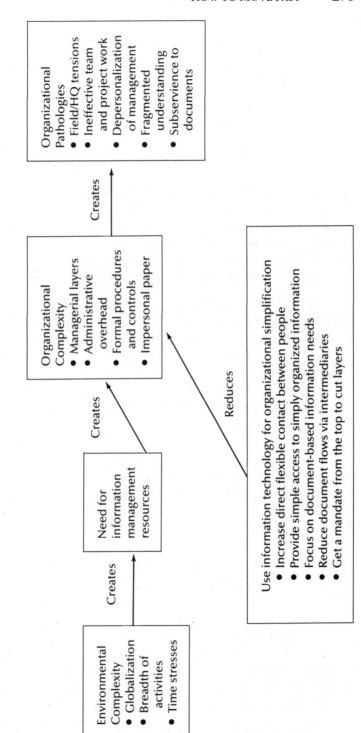

Environmental
Complexity
• Globalization
• Breadth of
  activities
• Time stresses

Creates

Need for
information
management
resources

Creates

Organizational
Complexity
• Managerial layers
• Administrative
  overhead
• Formal procedures
  and controls
• Impersonal paper

Creates

Organizational
Pathologies
• Field/HQ tensions
• Ineffective team
  and project work
• Depersonalization
  of management
• Fragmented
  understanding
• Subservience to
  documents

Reduces

Use information technology for organizational simplification
• Increase direct flexible contact between people
• Provide simple access to simply organized information
• Focus on document-based information needs
• Reduce document flows via intermediaries
• Get a mandate from the top to cut layers

ously to be expressed in organizational terms, not technical ones. Where the organizational end is clear, IT provides many means to achieving it.

Some Caveats in Defining a Practical Strategy

In defining a practical strategy for creating organizational simplicity, it is important to appreciate the linkages between cause and effect implicit in Figure 11–4.

1. Environmental complexity creates a need for additional information management resources. When those resources are based on adding people and procedures, they increase complexity, which creates organizational pathologies.
2. To reduce those pathologies, information resources must be added in a way that explicitly does not generate complexity.

This means that many efforts to apply traditional office technology and end-user computing will compound the problem. For example, an on-line planning and monitoring system too easily creates a powerful new set of information intermediaries, and generates its own administrative procedures—such as adding to field locations' workload in data collection and reporting—and inserts spreadsheets and analysis into an already overformal dialogue between corporate management and divisions.

Similarly, delayering without cutting out formal procedures and paper or cutting out paper while maintaining superfluous layers merely raises unmeetable expectations and discredits the argument that information technology in itself is a force for productivity. It is not. All the tools needed to use information technology to create organizational simplicity are widely available with videoconferencing as the main vehicle to repersonalize management, CD-ROM to provide simple access to mass information, EDI to improve transaction flows, and lap computers and videotex to link the field and head office.

THE FEDERATED ORGANIZATION STRUCTURE

Physical location and time have obviously always dominated organizational design. Matrix structures, decisions to centralize or decentralize, and reporting and control systems all reflect the problems of coordinating units across geography. Telecommunications expands options immensely and removes constraints on design. It makes dichotomy between centralization and decentralization entirely false.

The best analogy to the organizational structure and processes that telecommunications facilitates is the United States. At one level, the U.S. is highly decentralized—with states' rights and with different driving regulations, taxes, and school systems. At the same time, there are strong national policies, interstate highways, federal agencies, and so forth. At its most effective, this federated system balances local flexibility and diversity with central direction and coordination.

A comparable federated structure is emerging in a number of large organizations, though it is not yet a trend. It will and should be. The analogy to the United States is a powerful one that highlights the new fact that we can increase decentralization of operations and simultaneously increase centralization of coordination: that by building interstate highways we can still encourage or allow local diversity of regulations and operations. We do not have to use bureaucracy and control or to standardize procedures in order to coordinate across geography. Videoconferencing, through VSAT and fiber optics, will be an important element in facilitating the federated structure. "Bandwidth" will be less and less a constraint either on how "information" is moved or on what "information" means. The term will include conversations, meetings, documents, numbers, pictures, and the like. This allows:

1. *Distributed teams.* Teams that do not have to travel in order to communicate and that can be brought together quickly. Many of these teams will be in different organizations, working together to build the new electronic products opened up by strategic alliances across industries.

2. *Cooperative processing.* Several firms are using telecommunications to set up centers of excellence that can be shared electronically. For instance, one has consolidated its finance operations in New York. When it opened an office in Singapore, the managers there did not have to hire their own international tax expert or look for local help on investment strategies. They used New York—directly, constantly, and electronically. The firm is also consolidating its scientific computing in Ohio, exploiting economics of expertise that can be shared via cooperative processing.

3. *Location-independence.* Bechtel's visionary strategy is almost certain to be one of the exemplars of how to use telecommunications to transform the very idea of "organization" and "project-management." Bechtel will be able to assemble a team of skilled engineers "in" San Francisco and "send" them to, say, Saudi Arabia. The engineers will in fact be in Ireland and stay there. Bechtel is "off-sharing" project teams just as many firms offshore manufacturing. Bill Howard, the architect of this potential breakaway move to use telecommunications for organizational advantage, comments on how he was able to assemble a team to allow such a global project to occur: "The people who have been truly successful in this industry merge technology with business applications. They understand how to apply technology and they are able to explain it."

4. *Management and teamwork by conversation.* In the past years, there has been a flood of superb research on and development of software tools for "groupware," "group decision support," and "computer-augmented teams." They move the use of personal computers from stand-alone spreadsheet analysis and word processing to networked project management and commitment-oriented group work.

They go well beyond passive "communication" by electronic mail. For instance, COORDINATOR (Trademark Action Technology Inc.) provides the base for managing "conversations for action" and "conversation for possibility" that amounts to a new discipline and language for handling commitments and contracts—requests and promises—and that structures the entire process of ongoing, recurrent, and intermittent dialogs, alerting people to when promises are due, and tracking and retracking progress. COORDINATOR is perhaps the most striking and important tool since Lotus 1-2-3. The opportunities it provides to rethink "management" and "communication" make it—like 1-2-3—a powerful force for organizational redesign. Of course, it will be in no way easy to create the federated structure, restore organizational simplicity, manage location-independent teams, and manage by conversation. COORDINATOR, for instance, challenges people to think beyond vague communication and electronic mail messages whose intended action is unclear. It can generate quite strong counterattacks from people who see it as imposing a new electronic bureaucracy or who do not like making and keeping commitments but are experts in the art of organizational evasion.

Obviously, radical organizational change is always difficult to lead and manage. If a firm wants to exploit the tremendous emerging opportunities to use telecommunications for organizational advantage, its leaders must lead.

HOW CITIES COMPETE IN TIME

In the next decade, cities will compete through telecommunications just as they always have through other mechanisms relevant to maintaining a strong position in international trade and communication. In previous centuries, Bruges, Venice, Genoa, Lisbon, and Nuremberg, among many other European cities, were at the center of trading networks that gave them influence and wealth. Telecommunications merely changes the mechanisms by which such networks operate but certainly will have a significant impact on the relative strength of major cities worldwide. This has implications for both national policies on telecommunications, cities' strategies in competing, and firms' choice of locations and interest in those policies and strategies.

No city can be a really major business location in the world network without being a major money center; that requires electronic international banking and security trading. It must, too, invite rather than push away multinational corporations that are using computers and communications to evolve a federated structure; that requires a national policy of liberalization of telecommunications. It also needs to be a hub in the international airline network; that means the national "flag carrier" airline must have a strong electronic reservations and marketing base.

The new competitive game will be interesting to watch over the next decade. London must exploit its obvious advantage of Greenwich Mean Time, and fight off Amsterdam, which well recognizes the stakes, and watch Chicago. If deregulation of airlines comes to Europe, Amsterdam wants to ensure Schiphol is a major hub. British Airways will exploit Heathrow as a natural gateway to Europe. London attracts trading business in foreign exchange and securities from Tokyo, because of New York's disadvantageous time zone. Chicago is positioning as an alternative; the two hours from the East Coast makes a large competitive difference. It explains why the Chicago options market is open on Sunday evening.

Frankfurt will be a major airline hub, but the city's ability to compete has been badly impeded by Germany's consistent restrictive policies on telecommunications. The German deutschmark is the third currency of the world, along with dollars and yen, but the German financial centers are relatively weak in the global electronic network for the same reason. London trades close to twice as much in foreign exchange dealing as New York and Tokyo and close to 10 times that of Germany because of its time zone plus liberalized communications. Even though the French have been leaders in many aspects of telecommunications, the PTT's policy of tight control and almost prevention of multinationals' breadth of choice leaves Paris at a strong disadvantage; neither Charles de Gaulle airport nor the Paris Bourse figure to be major hubs in the world trade network.

Cities like Amsterdam, Toronto, Brussels, Chicago, and Singapore will benefit from time zone plus telecommunications. Hawaii, too, takes on a new strength. Physically isolated, so that a local trip to the mainland is 3,000 miles, it is electronically well placed as a center for satellite communications between not just the United States and the Far East but Europe through the United States to Asia.

More guesses can be made—obviously not predictions since telecommunications is an important enabler but not a direct causal factor and there are many, many other forces at work. India's strong potential as a resurgent force in the Asian economy requires it to strengthen its position as a trading center; satellite communications gives it a place in the electronic network. Miami will increase its role as both the electronic gateway to the Far East and a back up link from New York to London. For all its polarization of British society and highly criticized social polices, Margaret Thatcher's government may well have guaranteed London's presence as a major trading city in the world network, through its liberalization of telecommunications, the privatization of British Airways, and bold deregulation of the securities industry that well merited the description Big Bang.

All in all, telecommunications will play a more than incidental role in the growth and decline of cities in the international sphere.

CONCLUSION: THE MANAGEMENT AGENDA

This has been a book about business, not technology. This last chapter has addressed organizational issues, not technical ones. For over 25 years, firms have been trying, with increasing success, to harness information technology. Senior managers' role has been limited. They have mainly been approvers rather than initiators. They have helped things happen rather than made them happen.

Telecommunications, the highway system for integrated information technology, brings accelerating change. It brings organizational and business issues to the foreground of planning and action. It pushes senior managers to take on a very new relationship to what has traditionally been outside their scope. Several years ago, few executives would have defined understanding of telecommunications as part of the profile of the effective manager. If they do not do so now, they soon will.

Change is a problem, an opportunity, and a responsibility. As telecommunications stimulates radical business change and radical organizational change and in itself involves radical technical change, the problems and opportunities are bigger than before.

The managerial responsibility obviously is bigger, too. The sequence of innovation for telecommunications and business strategy depends on real management involvement and commitment. Ten years from now, most of the issues discussed in this book will be part of management common sense. For the next few years, many of them will be part of the management learning process.

There is an old Confucian curse: "May you live in interesting times." Telecommunications ushers in an intensely interesting time for all large firms. Telecommunications is a whole new arena where business imagination combined with understanding of just a few aspects of what the technology can do opens up entirely new ways of thinking about customers, markets, productivity, coordination, service, competition, products, and organization.

Telecommunications is about competition, innovation, risk, and uncertainty. That is an opportunity that all senior managers should welcome. The aim of this book has been to help prepare them to take it.

INDEX

ABOUT THE AUTHOR

Peter G.W. Keen is the executive director of the International Center for Information Technologies (ICIT).

Dr. Keen has been an advisor to the senior management of multinational corporations, airlines, and petrochemical and financial service firms in Europe, North and South America, and Asia. He has also advised governments and government agencies in several countries.

As an educator, he has served on the faculties of the Harvard Business School, the Massachusetts Institute of Technology, and Stanford University, and he was a visiting professor at the Wharton School of Business and the London Business School.

Before writing *Competing in Time: Using Telecommunications for Competitive Advantage*, Dr. Keen wrote the groundbreaking book *Decision Support Systems: An Organizational Perspective*. He has published extensively in journals such as *Barrons, Datamation, MIS Quarterly, Management Science*, and *Office: Technology and People*. In addition, Dr. Keen has appeared in a number of television programs on office automation and telecommunications.

Dr. Keen's research and practice have focused on the links between technology, business management, and organizational health; on effective implementation of new systems in the workplace; and on the critical policy decisions senior management must address to exploit the business opportunities of information technologies.

He received his undergraduate degree from Oxford University. His master's and doctoral degrees are from Harvard University.

The International Center for Information Technologies (ICIT) was founded by Dr. Keen in 1986, with funding by the MCI Communications Corporation, to help senior business and information service managers in large firms and public agencies plan for, manage, and understand information technologies.

From its Washington, D.C., and London, U.K., offices, ICIT sponsors research studies that address critical competitive and organizational issues. ICIT findings are translated into a variety of outreach programs and advisory services, including conferences, publications, strategic education courses, and videoconference programming.